CONTENTS

The Textbook iv
To the Student vi

Introduction: A Human Perspective 1

Part 1 Human Organization 6

1. Chemistry of Life 6
2. Cell Structure and Function 15
3. Introduction to Homeostasis 23

Part 2 Maintenance of the Body 31

4. Digestive System and Nutrition 31
5. Composition and Function of Blood 39
6. Cardiovascular System 46
7. Lymphatic System and Immunity 55
8. Respiratory System 64
9. Urinary System and Excretion 71

Part 3 Movement and Supportin Humans 79

10. Skeletal System 79
11. Muscular System 87

Part 4 Integration and Coordination in Humans 96

12. Nervous System 96
13. Senses 109
14. Endocrine System 117

Part 5 Reproduction in Humans 128

15. Reproductive System 128
16. Sexually Transmitted Diseases 136
17. Development and Aging 142

Part 6 Human Genetics 150

18. Chromosomal Inheritance 150
19. Genes and Medical Genetics 162
20. DNA and Biotechnology 169
21. Cancer 181

Part 7 Human Evolution and Ecology 188

22. Evolution 188
23. Ecosystems 196
24. Population Concerns 205

The Textbook

Human Biology, sixth edition, by Sylvia S. Mader has many student aids that you should be sure to use.

Text Introduction: The introductory chapter discusses the characteristics of humans and presents an overview of the book. It outlines the major biological principles and reviews the scientific method that allowed scientists to arrive at these principles.

Part Introduction: An introduction to each part highlights the central ideas of that part and specifically tells you how the topics within each part contribute to biological knowledge.

Chapter Concepts: Each chapter begins with a list of concepts that organize the content of the chapter into a few meaningful sentences. The concepts provide a framework for the content of the chapter.

The concepts are grouped under the major sections of the chapter and are page referenced for study. This numbering system, which is used in the text material and in the summaries, allows you to study the chapter in terms of the concepts presented.

Key Terms: Key terms are boldfaced in the chapter and are defined in context. These terms are page referenced in *Understanding the Terms* at the end of the chapter, and are defined in the glossary.

Internal Summary Statements: Short internal summary statements appear at the end of major sections and help you to focus your study efforts on the basics. Such statements are set off for easy identification.

Readings: Reading topics correlate to the subject matter of the chapter. Two types of readings are included in the text:

Health Focus readings review measures to keep healthy, such as the need for exercise ("Exercise, Exercise, Exercise", Chapter 10) and guidelines for preventing cancer ("Prevention of Cancer", Chapter 21).
Ecology Focus readings draw attention to a particular environmental problem, such as acid rain ("The Harm Done by Acid Deposition", Chapter 1).

Illustrations: The illustrations in *Human Biology* are consistent with multicultural educational goals.

Integrative illustrations relate micrographs with drawings. This enables students to see an actual structure alongside a diagram drawn similarly to the micrograph. Leaders and labels common to both the micrograph and the drawing allow the student to accurately locate and visualize a particular structure.

The *Visual Focus* illustrations are new to this edition. They illustrate in depth a main topic, system, or cycle in the chapter.

Bioethical Issues: This section is new to this edition. The Bioethical Issue topics are covered in a non-biased manner at the ends of the chapters. These topics stimulate thought and discussion about today's most relevant bioethical issues.

Chapter Summaries: The summary, called *Summarizing the Concepts,* is organized according to the major sections of the chapter, and the content of each section is summarized in a short paragraph or two. Chapter summaries offer a concise review of material in each chapter. You may read them before beginning the chapter to preview the topics of importance, and you may also use them to refresh your memory after you have a firm grasp of the concepts presented in each chapter.

Chapter Questions: *Studying the Concepts* are page-referenced short-answer questions that follow th[e] the material in the chapter. *Testing Your Knowledge of the Concepts* are objective questions that allow you to to answer recall-based questions. At least one question requires that you label a diagram or fill in a table. *Apply[ing] Knowledge to the Concepts* asks you to relate the concepts you have learned to matters of practical concern. Ans[wers to] *Testing Your Knowledge of the Concepts* and *Applying Your Knowledge to the Concepts* appear in Appendix A of [the] textbook.

Applying Technology to the Concepts: This section provides further resources on the Internet and on CD-RO[M for] studying the topics presented in the chapter.

Other

Further Readings: If you would like more information about a particular topic or are seeking references for a research paper, each part ends with a list of current articles and books to help you get started. Most entries are *Scientific American* articles and specialty books that expand on the topics covered in the chapter.

Appendices: The appendices contain optional information. Included is Appendix A, "Answer Key"; Appendix B, "Metric System"; Appendix C, "Table of Chemical Elements"; Appendix D, "Drugs of Abuse"; Appendix E, "Selected Types of Cancer"; and Appendix F, "Acronyms".

Glossary: The text glossary defines the terms most necessary for making the study of biology successful. With this tool, you can review the most frequently used terms. Pronunciations are given for words of greater difficulty.

Index: The text concludes with an index so that you may quickly locate the page or pages on which any given topic is discussed. Topics of special interest are highlighted in color.

To the Student

...was designed for the student and written to accompany your text, *Human Biology*. There is a study guide ...n corresponding text chapter. Each chapter contains the features listed here.

...dy Tips: The study tips help you plan the best way to study the chapter. Concepts that are easily confused and ...nology that is used for several purposes are also explained in this section.

Study Questions: The study questions are referenced to the sections of the text chapter and therefore can be answered as you study each section. They are also correlated to the study guide objectives. The study questions challenge you to read the appropriate section of the text and use this information to answer the questions. They provide you with the active practice you need to learn, understand, and apply the concepts in the text.

Definitions: Your study of human biology will likely require you to learn many new words. In fact, new terminology is often what frustrates students the most. To help you enjoy learning new terms, a Definitions Crossword or Definitions Wordsearch has been included at the end of the study questions in each chapter. Notice that each chapter in your textbook has an end-of-chapter glossary of selected key terms. The definitions used for the crossword and wordsearch puzzles are taken from the Understanding the Terms section. Study the terms in your textbook first, then use the crossword and wordsearch puzzles as a means of testing your knowledge.

Chapter Test: Most students recognize that there is a big difference between studying a chapter and taking a test on the chapter. This section is designed to help you overcome this hurdle. Taking this test just as you would an actual test will help you determine if you are ready to be tested by your instructor.

The chapter test is divided into two parts: the objective test and the thought questions. The objective test is a series of multiple-choice test questions (and perhaps some true/false questions) that are answered by simply selecting the correct answer. *The questions in the objective test are sequenced according to the objectives.* The thought questions require that you write out your answer in complete sentences. In many institutions, biology instructors are placing a new emphasis on developing critical thinking skills. This section will assist you in developing your writing and critical thinking skills.

Answer Key: Each chapter ends with an answer key, which contains answers for the study questions, definitions, and chapter test. Using the answer key can help you determine how well you are meeting the objectives of the chapter.

Introduction: A Human Perspective

STUDY TIPS

By entering into a study of human biology, you will learn about the structure and function of the body's systems. Therefore, you will be better equipped to make wise decisions about the health of yourself and your family members. You will also be learning about the evolution of humans and their place in the biosphere. Many human activities have impacted ecosystems for the worse, and your study should help you, to make choices that will help conserve ecosystems from now on.

This chapter introduces you to these topics, and it also describes the scientific method, by which scientists have discovered data that pertains to all aspects of human biology. As you study each chapter, read the integrated outline and then read the chapter, paying close attention to the boldface terms and summary statements. After each section, answer the questions at the end of the chapter for that section and answer the study guide questions for that section. Finally, take the chapter test in both the text and the study guide. Answers to Testing Your Knowledge in the text are found in Appendix A. Answers to all questions in the study guide are located at the end of each study guide chapter.

STUDY QUESTIONS

Study the text section by section as you answer the questions that follow.

1.1 Biologically Speaking (pages 2–7)

- Describe the biological characteristics that define human beings.
- Recognize how humans fit into the world of living things.
- Realize that some human activities threaten the biosphere.

1. The bodies of multicellular creatures, such as humans, are highly _____.
2. Humans are made up of cells, which are grouped into _____, and then organs.
3. When humans _____, they create a copy of themselves and perpetuate the species.
4. We are biological organisms and share traits with all other living things, but we are different from all others because of our cultural heritage. What is meant by "cultural heritage?"
   _____
   _____
5. Humans are the product of _____, an ongoing process that has occurred over millions of years.
6. _____, the group to which humans belong, have a nerve cord protected by a vertebral column.
7. Humans belong to the animal kingdom. List the other four kingdoms.
   a. _____ b. _____
   c. _____ d. _____
8. We are a part of the _____, the layer of life that surrounds the earth.
9. All organisms of the same species belong to a(n) _____ that inhabits a given area.
10. Plants store energy from the _____ in the form of organic compounds.
11. A large ecosystem is able to _____ its raw materials, like nitrogen and water, and needs only energy added from the outside.

12. By clearcutting tropical rain forests, we are causing the a._____ of many species of organisms.
    b. List other reasons why it is important to save tropical rain forests.
    _____
    _____

13. The number of organisms living in an area is referred to as _____.

1.2 THE PROCESS OF SCIENCE (PAGES 8–11)

- Understand how the scientific process is used to gather information and to arrive at conclusions.
- List the steps of the scientific method.

14. The _____ variable is the one being tested.
15. The _____ variable is the result or change that is observed in the experiment.
16. In any experiment, observations generate a._____, which can later be used to formulate a(n)
    b._____.
17. Referring to the following diagram, write the steps missing at
    a._____ b._____

```
         observations
              ↓
         previous data
              ↓
      formulation of
        hypothesis        ←─────┐
              ↓                 │
      observations              │
   and/or experimentation       │
              ↓                 │
         new data ──→ conclusion
                       1 ↓ 2 ↓ 3
                          theory
```

18. A unifying principle in biology is called a(n) _____.
19. The group in an experiment that is not receiving the treatment of interest is the _____ group.

1.3 SCIENCE AND SOCIAL RESPONSIBILITY (PAGE 11)

- Discuss how all people have a social responsibility to decide how to best use scientific knowledge for the benefit of all living things, including us.

20. Explain why every person has a social responsibility to ensure the proper use of scientific findings. _____
    _____
    _____

21. Science can sometimes provide technologies that are _____ disastrous, if not used properly.

DEFINITIONS CROSSWORD

Review key terms by completing this crossword puzzle, using the following alphabetized list of terms:

biodiversity
biosphere
cell
control
data
ecosystem
evolution
homeostasis
hypothesis
kingdom
reproduce
theory
vertebrates

Across
1. The total number of species within an ecosystem.
3. A major principle in biology, after much experimentation.
7. Community plus its nonliving habitat.
8. Species _____ to make offspring.
10. A theory in biology that suggests that all organisms arose from earlier forms.
11. What a scientist tests in an experiment.

Down
1. The layer of life surrounding this planet.
2. What the scientist gathers when making observations.
4. The _____ group receives no experimental treatment in an experiment.
5. Organ systems help the body maintain _____ or steady state.
6. Humans and other organisms with backbones are _____.
9. The highest group in our classification scheme.
12. The basic unit of life of all living things.

CHAPTER TEST

OBJECTIVE TEST

Do not refer to the text when taking this test.

_D_ 1. Choose the highest level of organization.
   a. cells
   b. tissues
   c. organs
   d. organ systems

_C_ 2. The basic unit of all living things is the
   a. nucleus.
   b. molecule.
   c. cell.
   d. atom.

_D_ 3. Organ systems of the body operate together to help maintain
   a. homeostasis.
   b. body temperature.
   c. reproductive status.
   d. circulation.

_B_ 4. Which organ system functions to convert food particles to nutrient molecules?
   a. nervous
   b. digestive
   c. respiratory
   d. immune

3

____ 5. Unlike other organisms on earth, humans have a
   a. large brain.
   b. vertebral column.
   c. manual dexterity.
   d. cultural heritage.

_b_ 6. To which of the following animals are humans most closely related?
   a. dogs
   b. chimpanzees
   c. wolves
   d. monkeys

_A_ 7. All of the members of the same species that occupy an area at the same time constitute a
   a. community.
   b. ecosystem.
   c. population.
   d. tribe.

____ 8. The variety of _____ between populations in an ecosystem tends to keep the system stable.
   a. predation
   b. commensalism
   c. interactions
   d. competition

____ 9. Ecosystems around the world are threatened more as
   a. human populations grow.
   b. the global temperature cools.
   c. insects become more numerous.
   d. seasons change.

____ 10. The process of science can best be described as
   a. objective.
   b. subjective.
   c. complex.
   d. hard to comprehend.

____ 11. Biologists use the scientific method to find out about nature. When observations are gathered, the next step is to
   a. design an experiment.
   b. consult a statistician.
   c. formulate a hypothesis.
   d. come up with a theory.

____ 12. The information from a scientific experiment is best presented
   a. mathematically and graphically.
   b. concisely.
   c. succinctly.
   d. subjectively.

____ 13. A new medication is being tested to determine its effectiveness in controlling obesity in white laboratory mice. A suitable control group would be
   a. any group of mice, caught from the wild.
   b. a group of white laboratory mice, kept in the same way, not given the medication.
   c. a group of white laboratory mice given a lower dosage of the medication.
   d. No control is needed, just see whether the mice lost weight.

____ 14. The principle of evolution is a
   a. hypothesis.
   b. unproven observation.
   c. educated guess.
   d. theory.

____ 15. In order for scientific information to be used properly, for the benefit of all living things, we must all be
   a. literate.
   b. smart.
   c. subjective.
   d. socially responsible.

THOUGHT QUESTIONS

Answer in complete sentences.

16. Why is it preferred to have mathematical data gathered in an experiment?

17. Why is biodiversity important in ecosystems, and why are scientists alarmed at the current rate of loss of biodiversity?

18. Give an example of an area that cannot be subjected to the scientific method.

Test Results: _____ number correct ÷ 18 = _____ × 100 = _____ %

Answer Key

Study Questions

1. organized **2.** tissues **3.** reproduce **4.** Humans learn from their parents and others how to act in a socially acceptable, civilized manner, which sometimes sets us apart from the rest of nature. **5.** evolution **6.** Vertebrates **7. a.** plants **b.** fungi **c.** protists **d.** monerans **8.** biosphere **9.** population **10.** sun **11.** cycle **12. a.** extinction **b.** Aside from preventing the extinction of many fascinating species, it is important to save tropical rain forests because of the carbon they store (releasing this carbon contributes to global warming), and because of the potential for new medicines that may be found in organisms residing there. **13.** biodiversity **14.** experimental **15.** dependent **16. a.** data **b.** hypothesis **17. a.** formulation of hypothesis **b.** conclusion **18.** theory **19.** control **20.** Ethical and moral judgments must be made by all people. This includes the area of the use of scientific findings. Each of us plays an important role within the ecosystems of this planet; it is up to us to see that the remainder of the organisms here, including us, are protected from abuses. **21.** ecologically

Definitions Crossword

Across:
1. biodiversity **3.** theory **7.** ecosystem **8.** reproduce **10.** evolution **11.** hypothesis

Down:
1. biosphere **2.** data **4.** control **5.** homeostasis **6.** vertebrates **9.** kingdom **12.** cell

Chapter Test

1. d **2.** c **3.** a **4.** b **5.** d **6.** b **7.** c **8.** c **9.** a **10.** a **11.** c **12.** a **13.** b **14.** d **15.** d **16.** Mathematical data can be subjected to statistical analysis; therefore it is objective and not as likely to be subject to individual bias. **17.** Biodiversity helps to ensure that resources are divided up fully and provides structure and stability to ecosystems. The current rapid loss of biodiversity is due to the activities of humans. If loss of biodiversity continues at its present rate, many ecosystems will no longer function as they once did. This could threaten the very existence of the biosphere and certainly the survival of humans. **18.** Science cannot prove the existence of God, for example.

PART I HUMAN ORGANIZATION

1

CHEMISTRY OF LIFE

STUDY TIPS

It is sometimes difficult for a beginning biology student to appreciate chemistry, but keep in mind that understanding basic chemistry will help you to understand the workings of the cell, the subject of the next chapter.

The term *nucleus,* as it is applied to atomic structure (p. 16), is not the same "nucleus" that stores genetic information inside the cell. Just as there are different usages of the term *center* (in the middle; large building for performing arts; or football position), there are several ways the term *nucleus* is used in biology.

You can predict the number of electrons for a given atom by looking at its number of protons (p. 17). Normally, atoms are electrically neutral, which means the number of positive charges (protons) balances the number of negative ones (electrons). Take time to fully comprehend the basic structure of an atom. It will help you understand how bonds form (pp. 19–20) between atoms and how larger biological molecules (p. 26–37) are constructed.

STUDY QUESTIONS

Study the text section by section as you answer the questions that follow.

1.1 ELEMENTS AND ATOMS (PAGES 16–17)

- Indicate the relationship between elements and atoms.
- Describe the structure of an atom and state the location and electrical charge of protons, neutrons, and electrons.
- Be able to describe an electrically neutral atom.
- Describe a stable, unreactive atom.
- Explain the nature of isotopes and radioactive isotopes.

1. The smallest functional unit of an element is a(n) _____.
2. List three examples of elements.
   a._____ b._____ c._____
3. In this drawing of an atom, write *P* for protons, *N* for neutrons, and *E* for electrons beside *a* and *b*. Indicate the electrical charge carried by each type of particle.
   a._____
   b._____

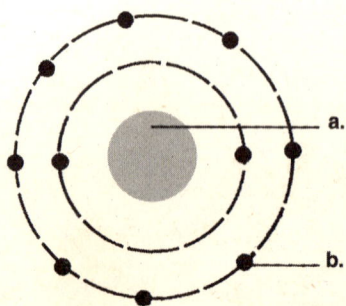

4. An atom is said to be electrically neutral when it has equal numbers of a._____ and b._____.
5. Stable, unreactive atoms usually have _____ electrons in their outermost shells.
6. The atomic number of an atom is equal to its number of _____.
7. The atomic weight of an atom is equal to its number of a._____ and b._____.
8. _____ of an element are atoms that differ in their number of neutrons.
9. Atoms of _____ isotopes, such as carbon-14, radiate high-energy particles from the nucleus as they decay to carbon-12.

1.2 MOLECULES AND COMPOUNDS (PAGES 19–20)

- Describe the nature of a molecule.
- Be able to describe what happens to the electrons in an ionic bond.
- Understand the nature of covalent bonds.

10. A(n) _____ forms when two or more atoms are bonded together.
11. Regular table salt is sodium chloride, NaCl. Sodium has 11 protons. How many electrons will it have when it is electrically neutral? a._____ How many electrons will be in its outermost shell? b._____
12. Chlorine has 17 protons. How many electrons will chlorine have when it is electrically neutral? a._____ How many electrons are in its outermost shell? b._____
13. Based on your answers for questions 11 and 12, what type of bond do you expect will form between sodium and chlorine? _____
14. Why? _____
_____
_____
_____
15. When pairs of electrons are shared between atoms, _____ bonds result.
16. In a triple bond, _____ pairs of electrons are shared.

1.3 WATER AND LIVING THINGS (PAGES 21–25)

- List the important properties of water molecules.
- Discuss the general chemical properties of acids and bases.
- Explain and be able to use the pH scale.

17. One type of inorganic molecule, _____, makes up 60–70% of all living things.
18. What type of bond occurs *within* water molecules? _____
19. What type of bond occurs *between* the water molecules? _____
20. The following diagram illustrates the dissociation of water molecules.

$$H–O–H \rightleftarrows H^+ + OH^-$$
water   hydrogen   hydroxide
              ion   ion

The pH scale is based on the relative abundances of hydrogen ions and hydroxyl ions.

When there is more H^+ and less OH^-, is the pH acidic or basic? a._____

When there is less H^+ and more OH^-, is the pH acidic or basic? b._____

21. Technically speaking, a(n) a._____ gives off hydrogen ions in a solution, while a(n) b._____ takes up hydrogen ions or gives off hydroxyl ions.
22. Because of hydrogen bonding, water sticks to itself and is called a._____. Is this property due to hydrogen bonding or to polarity of the water molecule? b._____
23. A(n) _____ prevents rapid changes in pH by taking up either hydrogen or hydroxide ions.

1.4 MOLECULES OF LIFE (PAGE 26)
- Describe the basic composition of organic molecules.

24. Carbon shares its electrons with as many as _____ different atoms at once.
25. Organic molecules can often attain large sizes and are thus called _____.
26. Organic molecules, like the one illustrated here, usually have a "backbone" made out of the element _____.
27. a._____ and b._____ are also elements found in most organic molecules.
28. Are organic molecules characteristic of living or nonliving matter? _____

29. Large organic molecules are built of smaller units, called _____.

1.5 CARBOHYDRATES (PAGES 27–28)
- Explain the formation of polymers by condensation synthesis and their breakdown by hydrolysis.
- Describe the function and structure of carbohydrates.

30. Large organic molecules are built through a reaction called a._____ _____ that removes a molecule of b._____.
31. During digestion, organic molecules are broken down into component parts by adding a._____ molecules through a reaction referred to as b._____.
32. The building blocks of carbohydrates are the simple sugars, known technically as a._____. The simple sugar b._____ is one example.
33. Carbohydrates are the primary source of _____ in the diet.
34. When many glucose molecules are joined, the macromolecule _____ results.
35. Table sugar, sucrose, is an example of a(n) _____, being built from two sugar molecules.
36. Humans store glucose in the form of a polymer known as a._____ in their livers. Another polysaccharide, b._____, lends structural support to plant cell walls and is not digestible by humans.

1.6 Lipids (pages 29–30)

- Describe the function and the structure of lipids, soaps, phospholipids, and steroids.
- Recognize the difference between saturated and unsaturated fatty acids.

37. Dietary fats are most commonly in the form of a._____, which are formed from one b._____ molecule and three c._____ molecules. Since they repel water and are nonpolar, they are called d._____ fats.

38. Lipids function as a._____ and keep us warm; they provide long-term storage of b._____; and they form a protective c._____ around our internal organs.

39. Soaps break up fats through a process called a._____ in which soap molecules surround individual fat molecules. b._____ from the gallbladder functions in the same way in the digestive system.

40. A type of lipid with a polar phosphate group, called a(n) a._____, is a major component of plasma membranes.

41. Cholesterol is the precursor of several lipids, known as a._____, that function as hormones in the body.

42. Write the words *saturated* and *unsaturated* on the line below the appropriate structure.

a. _____   b. _____

43. What does saturated mean? _____
_____

1.7 Proteins (pages 31–33)

- Describe the functions and basic structure of proteins.
- Compare the primary, secondary, tertiary, and quaternary structure of proteins.

44. The building blocks (monomers) of proteins are the _____.

45. In the following diagram of an amino acid, —NH₂ is the a._____ group.
—COOH is the b._____ group.

c. How does a peptide bond form between two adjacent amino acids?
_____
_____
_____

46. Certain proteins can function as a._____ to catalyze chemical reactions in the cell. Fingernails contain another important protein called b._____. Other proteins, such as c._____ and d._____, are important structural components of connective tissue. The e._____ proteins in blood help transport molecules, fight infections, and help blood to clot.

9

47. a._____ structure in proteins refers to the sequence of amino acids. b._____ structure is any twisting of the chain of amino acids. Folding up of the chain is considered to be c._____ structure. d._____ structure exists when two or more chains are folded together, as in the hemoglobin molecule.

48. When proteins are exposed to heat, chemicals, or extremes in pH, they undergo a change in shape. They no longer function as usual, so we say they have been _____.

1.8 NUCLEIC ACIDS (PAGES 34–35)

- Discuss the structure of nucleic acids, and describe their functions in cells.
- Describe the structure and function of ATP.

49. The building blocks (monomers) of nucleic acids are _____.

50. The nucleic acid a._____ stores the genetic information for the cell. Another nucleic acid, b._____, works with DNA to specify the order of amino acids during the synthesis of proteins.

51. Refer to the following diagram of a strand of nucleotides to answer questions a–d.
    a. What molecule is represented by S? _____
    b. What molecule is represented by B? _____
    c. How many different types of B are in DNA? _____
    d. What type of bond is represented by the lines? _____

52. One nucleotide, a._____, functions as an energy carrier inside cells. Energy is released from this molecule when a(n) b._____ group is removed.

53. Associate the molecules in question 34 with this diagram.

    a. What monomers should be associated with (1) in the diagram?
    _____

    b. What macromolecules should be associated with (2)? _____
    _____

54. Complete this summary table.

Macromolecule	Monomer	Function
Proteins		
Carbohydrates, e.g., glycogen		
Lipids		
Nucleic acids		

55. Of the macromolecules listed in the table in question 54,

   Which are *most* concerned with energy? a._____

   Which one forms enzymes? b._____

   Which one makes up genes? c._____

DEFINITIONS CROSSWORD

Review key terms by completing this crossword puzzle using the following alphabetized list of terms:

atom
ATP
base
carbohydrate
cellulose
DNA
electron
glucose
ionic
isotope
lipid
nucleotide
protein

Across
1. Complex polysaccharide found in plant tissue that we cannot digest.
3. Organic compound made up of long chains of amino acids.
5. Type of bond formed between oppositely charged ions.
6. Monosaccharide that serves as a building block for starch.
7. A negatively charged subatomic particle.
8. Nucleotide that serves as the energy carrier for the cell.
11. A nucleic acid that serves as the cell's genetic material.
12. An atom that differs in its number of neutrons.

Down
1. Organic compound type that includes mono- and polysaccharides.
2. Insoluble organic compound, such as fats.
4. Building block of nucleic acid.
9. Molecule that releases OH⁻ in solution.
10. Functional unit of an element.

Chapter Test

Objective Test

Do not refer to the text when taking this test.

__A__ 1. The atomic weight of an atom is determined by the number of
   a. protons.
   b. neutrons.
   c. electrons.
   d. protons and neutrons.

____ 2. Isotopes differ due to the number of
   a. protons.
   b. neutrons.
   c. electrons.
   d. protons and neutrons.

____ 3. When an atom either gains or loses an electron, it becomes
   a. electrically neutral.
   b. an ion.
   c. stable.
   d. unreactive.

____ 4. Chlorine has 17 protons. When chlorine becomes the chloride ion (Cl⁻), it has
   a. gained an electron.
   b. lost an electron.
   c. gained a proton.
   d. lost a neutron.

____ 5. When sodium interacts with chlorine, sodium loses an electron while chlorine gains one. This interaction forms
   a. an ionic bond.
   b. a condensation synthesis.
   c. a condensation.
   d. a covalent bond.

____ 6. Bonds between carbon and hydrogen or oxygen and hydrogen are generally
   a. hydrogen bonds.
   b. ionic bonds.
   c. covalent bonds.
   d. weak and highly transient.

____ 7. When a strong acid, such as hydrochloric acid, is added to water,
   a. hydrogen ions are taken up.
   b. hydroxide ions are released.
   c. hydrogen ions are released.
   d. the pH stays the same.

____ 8. A pH of 11.5 is considered to be
   a. slightly acidic.
   b. strongly acidic.
   c. strongly basic.
   d. about neutral.

____ 9. Which of these is true of organic molecules?
   a. usually ionic bonding
   b. always contain carbon and hydrogen
   c. mainly associated with nonliving matter
   d. are found only in organisms with organ systems

__A__ 10. Polymers are
   a. chains of building-block molecules.
   b. formed by dehydration.
   c. broken down by hydrolysis.
   d. All of these are correct.

__D__ 11. Which pair of molecules is mismatched?
   a. amino acid–protein
   b. fatty acid–lipid
   c. glucose–starch
   d. glycerol–nucleic acid

For questions 12–16, match the following answers for lipids to one of the statements that follows.
   a. triglyceride   b. unsaturated fatty acid
   c. saturated fatty acid   d. phospholipid
   e. soap

____ 12. Not truly a lipid, this substance helps to emulsify fats.

____ 13. Has one or more double bonds along the fatty acid chain.

____ 14. Made up of a glycerol and three fatty acid molecules.

____ 15. Has hydrogen at every position along the fatty acid chain.

____ 16. A major component of cell membranes.

____ 17. Which molecule is unsaturated?

____ 18. The primary structure of a protein refers to its
   a. three-dimensional shape.
   b. order of amino acids.
   c. order of nucleic acids.
   d. orientation of the amino acids.

____ 19. Proteins are polymers of _____, which sometimes function to _____.
   a. amino acids, catalyze chemical reactions
   b. glucose, build muscle strength
   c. nucleotides, synthesize proteins
   d. ribosomes, produce quick energy

___20. In any amino acid, —NH₂ is the _____ group, and —COOH is the _____ group.
   a. carboxyl; amino
   b. amino; carboxyl
   c. peptide; fatty acid
   d. hydroxyl; carboxyl

___21. A bond forming between the —NH₂ group of one amino acid and the —COOH group of another amino acid is called a
   a. double bond.
   b. ionic bond.
   c. hydrogen bond.
   d. peptide bond.

___22. Twisting of a chain of amino acids into an α-helix is termed the _____ structure of a protein.
   a. primary
   b. secondary
   c. tertiary
   d. quaternary

For questions 23–26, match the four terms to the statements below.
   a. RNA   b. DNA   c. nucleotide   d. ATP

___23. Building block of the nucleic acids.
___24. Functions with DNA in the process of protein synthesis.
___25. The genetic material of the cell.
___26. The energy currency of the cell.

THOUGHT QUESTIONS

Answer in complete sentences.

27. Give one example of how radioactive isotopes are used in research or medicine.

28. Describe two ways in which the properties of water allow the existence of life.

Test Results: _____ number correct ÷ 28 = _____ x 100 = _____%

ANSWER KEY

STUDY QUESTIONS

1. atom   2. a. carbon b. hydrogen c. oxygen   3. a. protons (positive charge), neutrons (no charge) b. electrons (negative charge)   4. a. protons b. electrons   5. eight   6. protons   7. protons, neutrons   8. Isotopes   9. radioactive   10. molecule   11. a. eleven b. one   12. a. seventeen b. seven   13. ionic   14. Once sodium gives up the electron in its outermost shell, it is a stable atom. Chlorine also becomes stable by accepting the electron from sodium, now with eight electrons in its outermost shell. Since the chlorine atom now bears a net negative charge, and sodium a net positive charge, the two atoms (now ions) are attracted to each other and held together by an ionic bond.   15. covalent   16. three   17. water   18. polar covalent bond   19. hydrogen bond   20. a. acidic b. basic   21. a. acid b. base   22. a. cohesive b. hydrogen bonding   23. buffer   24. four   25. macromolecules   26. carbon   27. a. Hydrogen b. oxygen   28. living   29. monomers   30. a. condensation synthesis b. water   31. a. water b. hydrolysis   32. a. monosaccharides b. glucose   33. energy   34. polysaccharide   35. disaccharide   36. a. glycogen b. cellulose   37. a. triglycerides b. glycerol c. fatty acid d. neutral   38. a. insulation b. energy c. padding   39. a. emulsification b. Bile   40. phospholipid   41. steroids   42. a. unsaturated b. saturated   43. The carbon chain has all the hydrogens it can hold; there are no double bonds between carbon atoms.   44. amino acids   45. a. amino b. carboxyl c. When the carboxyl (acidic) group of one amino acid reacts with the amino group of another amino acid, a molecule of water is removed, and the peptide bond forms.   46. a. enzymes b. keratin c. collagen d. elastin e. globular   47. a. Primary b. secondary c. tertiary d. Quaternary   48. denatured   49. nucleotides   50. a. DNA b. RNA   51. a. sugar b. base c. four d. hydrogen   52. a. ATP b. phosphate   53. a. glucose, amino acids, glycerol and fatty acids, nucleotides b. fat, polysaccharide, polypeptide, nucleic acid   54. See Table 1.2, p. 35 in the text.   55. a. carbohydrates and lipids b. protein c. nucleic acids

13

Definitions Crossword

Across:
1. cellulose 3. protein 5. ionic 6. glucose 7. electron 8. ATP 11. DNA 12. isotope

Down:
1. carbohydrate 2. lipid 4. nucleotide 9. base 10. atom

Chapter Test

1. d 2. b 3. b 4. a 5. a 6. c 7. c 8. c 9. b 10. d 11. d 12. e 13. b 14. a 15. c 16. d 17. b 18. b 19. a 20. b 21. d 22. b 23. c 24. a 25. b 26. d 27. Radioactive isotopes are used to determine the age of fossils. They are also used in many diagnostic medical tests. 28. Because ice is less dense than water, bodies of water freeze from the top down. Aquatic organisms can survive throughout the winter in the cold water beneath the ice. The cohesiveness of water enables water to travel easily through a tube, such as it does within blood vessels. The water molecules adhere to one another, forming a continuous column of fluid.

2
CELL STRUCTURE AND FUNCTION

STUDY TIPS

As you study this chapter, remember that the cell is the basic unit of *all* living things. Even though we are complex multicellular creatures with many organ systems, it is our *cells* that must receive nourishment and oxygen in order to survive. When enough of our cells die, so do we.

Spend time studying the structure and function of the plasma membrane and cellular organization (pp. 42–52). In the chapters that follow, you will examine our various organ systems. If you understand the functions of the cell, it will aid your understanding of these organ systems.

STUDY QUESTIONS

Study the text section by section as you answer the questions that follow.

2.1 CELL SIZE (PAGES 42–43)

- Be able to explain why cells are small by nature.
- Recognize that different types of microscopy reveal different features of the cell.

1. Cells are small to enable them to maintain a favorable a._____ to b._____ ratio. This helps the diffusion of nutrients and gases throughout the cell.

2. As a student, you are most likely to encounter the a._____ microscope in the laboratory. The b._____ electron microscope scans the surfaces of samples to look at surface features. The c._____ electron microscope is capable of viewing the internal features of cells. Which type of microscope has the best resolving power? d._____

2.2 CELLULAR ORGANIZATION (PAGES 44–54)

- Describe the structure of the plasma membrane.
- Explain the variety of methods used by cells to allow passage of materials through the plasma membrane.
- Predict the results of placing a cell in solutions of various tonicity.
- Describe the structure of the nucleus and its importance to the cell.
- Discuss the features and importance of ribosomes.
- Explain the relationship between the endoplasmic reticulum, the Golgi apparatus, and lysosomes.
- Explain the structure and function of the mitochondria within the cell.
- Describe the structure of the cytoskeleton within the cytoplasm.
- Describe the relationship between centrioles and cilia or flagella.

3. Using the list of terms that follows, label the parts of the plasma membrane indicated in this diagram.
    carbohydrate chain
    cholesterol
    cytoskeleton filaments
    glycolipid
    glycoprotein
    phospholipid bilayer
    protein molecule

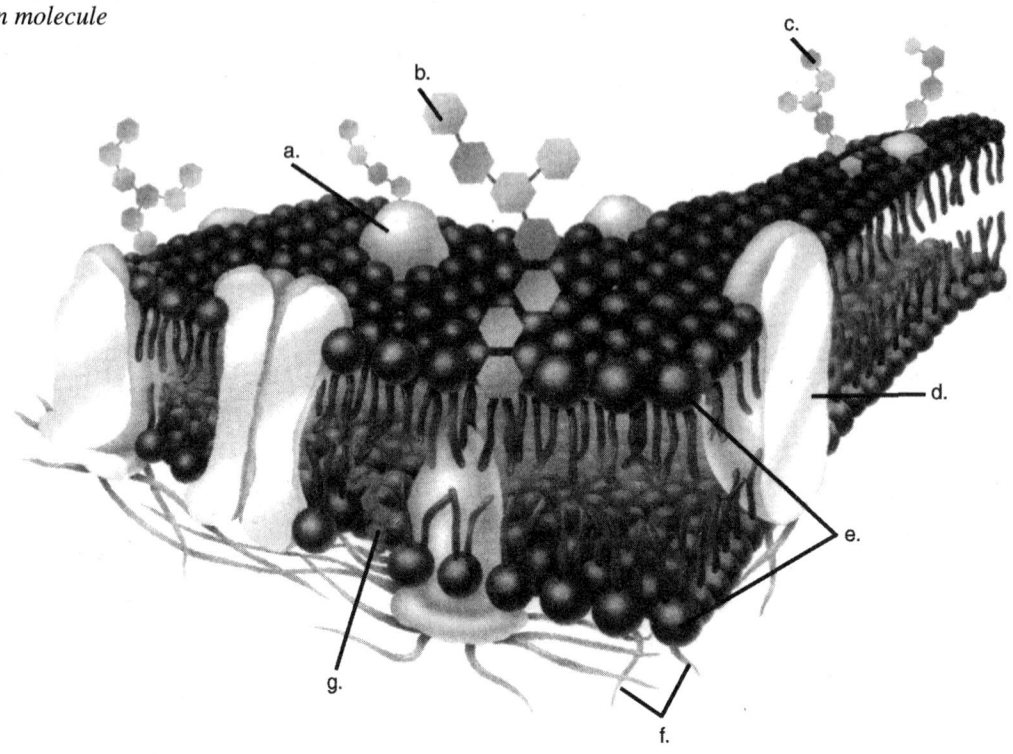

4. The plasma membrane is selectively permeable because it _____ what substances cross through the membrane.

5. Within the bilayer, a._____, or hydrophilic, heads extend outward and nonpolar, or b._____, fatty acid tails extend to the interior of the membrane.

6. a._____ is the passive movement of molecules from an area of greater concentration to an area of lower concentration. b._____ is a special case of this same process, involving the movement of water across a membrane.

7. If a cell is placed in a(n) a._____ environment, it neither gains nor loses water. When a cell enters a(n) b._____ solution, it gains water until it bursts.

8. a._____ transport involves a carrier protein assisting a larger molecule across the membrane with no expenditure of energy. b._____ transport carries materials across membranes against a concentration gradient and requires energy.

9. During the process known as a._____, the plasma membrane pinches off a vesicle around a particle the cell is engulfing. b._____ is the reverse process, during which the cell expels contents to the outside.

10. Using the blanks provided and the list of terms that follows, name the cellular structures indicated in the drawing and state the function of each.

centriole *nucleus*
Golgi apparatus *ribosome*
lysosome *rough ER*
microtubule *smooth ER*
mitochondrion *vacuole*
nucleolus

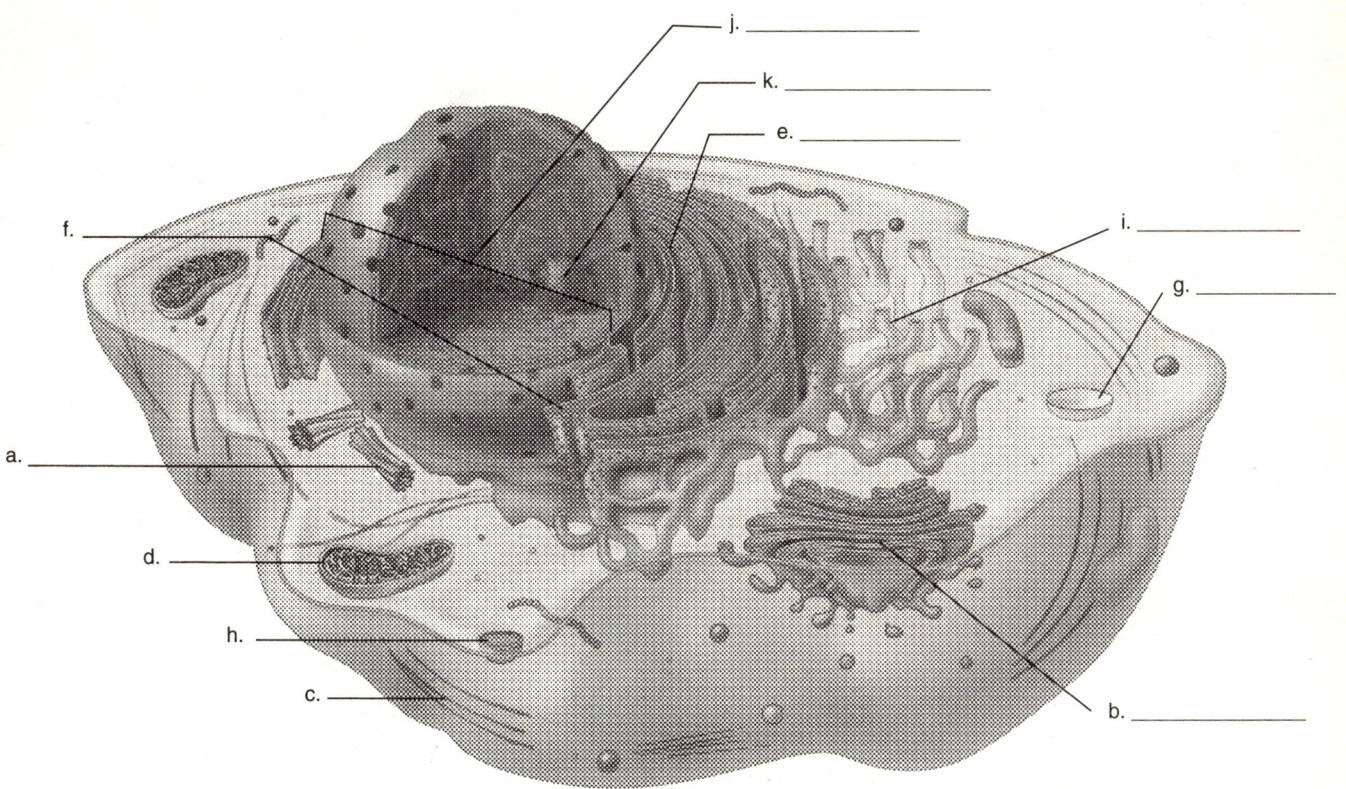

Name **Function**

a. _____ _____
b. _____ _____
c. _____ _____
d. _____ _____
e. _____ _____
f. _____ _____
g. _____ _____
h. _____ _____
i. _____ _____
j. _____ _____
k. _____ _____

11. The nucleus is enclosed by the a._____, which contains b._____ through which materials can pass into the cytoplasm. Within the nucleus, the c._____ is the site of production of rRNA for the manufacture of d._____.

12. Ribosomes are associated with a._____ endoplasmic reticulum, where proteins are made. Proteins travel in vesicles to the b._____, which packages them for secretion.

13. The cytoskeleton has two types of filaments of interest to us, a._____ and b._____. Centrioles give rise to c._____, which direct the formation of d._____ and e._____.

2.3 Cellular Metabolism (pages 54–57)

- Understand the functioning of a metabolic pathway and the importance of enzymes within it.
- Describe a generalized equation for an enzymatic reaction.
- Explain the function of coenzymes.
- Understand the activities that occur during aerobic cellular respiration, and relate the results to the three metabolic pathways involved.
- Associate mitochondria with the Krebs cycle and the electron transport system.
- Understand what conditions are needed for the process of fermentation compared to those of aerobic cellular respiration.

14. A a._____ _____ is a series of b._____-catalyzed reactions that must occur in a particular sequence.

15. The interaction of an enzyme with its target molecule can be summarized as follows: E + S —> a._____ —> E + P, where E stands for b._____, S for c._____, and P for d._____.

16. Nonprotein organic molecules that sometimes assist the functioning of enzymes are called a._____. b._____ is one example.

17. Label the matrix and cristae in this drawing of a mitochondrion.

Which of these do you associate with the Krebs cycle? c._____ Which of these do you associate with the electron transport system? d._____ Where does glycolysis take place? e._____

18. Refer to the following diagram and answer the questions that follow.

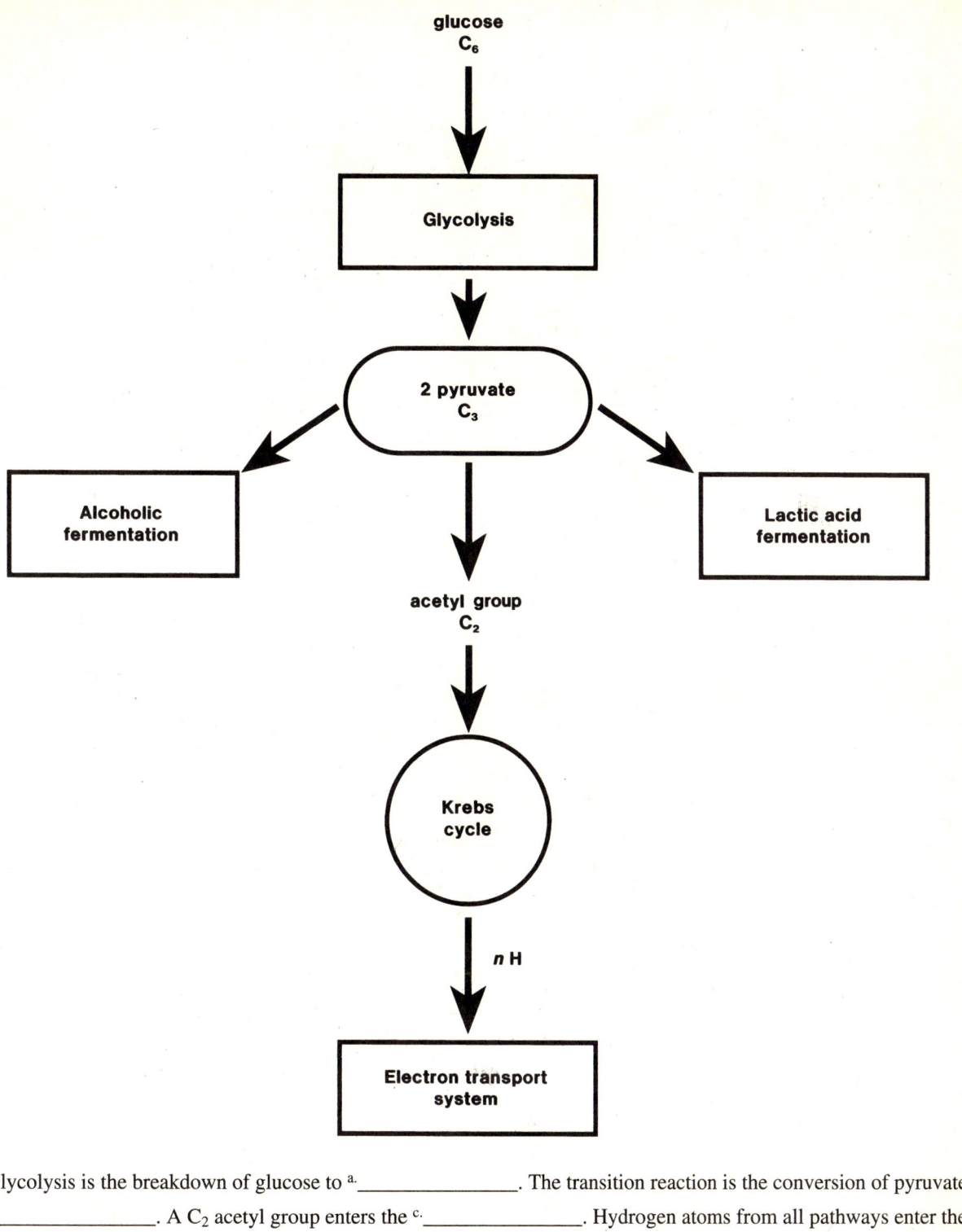

Glycolysis is the breakdown of glucose to a._____. The transition reaction is the conversion of pyruvate to b._____. A C₂ acetyl group enters the c._____. Hydrogen atoms from all pathways enter the d._____.

19. Which portion of cellular respiration generates the greatest amount of ATP? _____

20. If no oxygen is available in the cell, pyruvate is converted to a._____ via a pathway known as b._____.

19

DEFINITIONS WORDSEARCH

Rewrite key terms by completing this wordsearch, using the following alphabetized list of terms:

```
S E E T H E C A T R U N T T H I M O
E M O S O M O R H C B R P F A M I P
T H E R I M O T R Y C Y L T T O C K
T H E R I O T B G T D O A G D H R E
G I N N Y O C Y T O S K E L E T O N
R E N T O F R D C P T Y I O S H T I
A Y L I L L K A Q L W Z E U P P U P
F U N I D A R Y T A Y V L K C I B L
R E T N U G G U N S L O O H G F U M
L I T C O E N Z Y M E U I U O D L J
K I L Y T L N U T L R U R O M M E E
F L A G G L T M C I Y O T R E E S I
G O L G A U B U S O K J N O P L Y G
C I L I U M N R T E L L E N A G R O
N U C L A I U B F I T O C J U M P G
```

centriole
chromosome
cilium
coenzyme
cytoplasm
cytoskeleton
flagellum
microtubule
nucleolus
organelle

a. _____ Cylinder containing microtubules in a 9 + 0 arrangement and associated with the formation of basal bodies.
b. _____ Rodlike structure in the nucleus; DNA.
c. _____ Short, hairlike extension from the cell.
d. _____ Slender, long extension from the cell.
e. _____ Structure within the cell, surrounded by a membrane, with a specific function.
f. _____ A region inside the nucleus where rRNA is produced for the manufacture of ribosomes.
g. _____ Cytoskeleton component made up of the protein tubulin.
h. _____ Nonprotein organic molecule that aids the action of an enzyme.
i. _____ A network of protein filaments found within the cell. Its function is to move organelles around, among other things.
j. _____ Semifluid medium outside the nucleus in the cell.

CHAPTER TEST

OBJECTIVE TEST

 1. Cells are small so that
   a. they require fewer nutrients.
   b. they can reproduce rapidly.
   c. materials can easily diffuse throughout the cell.
   d. they can more readily become components of tissues.

_____ 2. Which type of microscopy has the best resolving power?
   a. simple compound microscope
   b. scanning electron microscope
   c. transmission electron microscope
   d. Both electron microscopes have the same resolving power.

_____ 3. Which type of molecule forms a bilayer within the plasma membrane?
   a. carbohydrate
   b. protein
   c. phospholipid
   d. nucleic acid

 4. A small lipid-soluble molecule readily passes through the plasma membrane. Which is the most likely explanation?
   a. A carrier protein must be carrying out facilitated transport.
   b. Diffusion occurs, since the membrane is made up of lipid molecules.
   c. The cell is expending energy to do work.
   d. Endocytosis occurs as the membrane engulfs a particle.

__A__ 5. Which of these does not require an expenditure of energy?
   a. diffusion
   b. osmosis
   c. facilitated transport
   d. None of these require energy.

____ 6. A large water-soluble molecule is moved from an area of greater concentration outside to one of lower concentration inside the cell. This is possible because of which explanation?
   a. A carrier protein must be carrying out facilitated transport.
   b. Diffusion occurs since the membrane is made up of lipid molecules.
   c. The cell is expending energy to do work.
   d. Endocytosis occurs as the membrane engulfs a particle.

____ 7. The thyroid gland contains a higher concentration of iodine than the blood of the body. This is most likely due to
   a. diffusion.
   b. active transport.
   c. facilitated transport.
   d. endocytosis.

__D__ 8. An animal cell always takes in water when it is placed in a(n)
   a. hypertonic solution.
   b. osmotic solution.
   c. isotonic solution.
   d. hypotonic solution.

__C__ 9. The nucleus is considered the control center of the cell because it
   a. is located in the exact center of the cell.
   b. houses the nucleolus.
   c. contains the genetic information of the cell.
   d. has nuclear pores through which materials can pass.

____ 10. Which of these cellular structures is NOT composed of membranes?
   a. ribosomes
   b. Golgi apparatus
   c. lysosomes
   d. the endoplasmic reticulum

____ 11. Which organelle has a 9 + 0 arrangement of microtubules and is associated with the formation of basal bodies?
   a. actin filaments
   b. spindle fibers
   c. flagellum
   d. centriole

____ 12. Cilia and flagella contain
   a. centrioles.
   b. microtubules.
   c. basal bodies.
   d. actin filaments.

____ 13. The cytoskeleton of the cell is composed of
   a. protein.
   b. actin filaments.
   c. microtubules.
   d. All of these are correct.

For questions 14 through 19, match the functions with the proper cellular structure.
   a. mitochondria   b. ribosomes   c. lysosomes
   d. centrioles   e. Golgi apparatus   f. cilia

__E__ 14. packaging and secretion
____ 15. hydrolytic enzymes
____ 16. energy production
____ 17. protein synthesis
__A__ 18. cell division
____ 19. cell locomotion
____ 20. NAD performs what function in cellular respiration?
   a. It is an oxygen carrier.
   b. It carries hydrogen atoms.
   c. It functions as an enzyme.
   d. It facilitates the transition reaction.

__A__ 21. In a metabolic pathway, each enzyme catalyzes a specific reaction in
   a. sequence.
   b. random order.
   c. organelles only.
   d. anaerobic conditions.

____ 22. The pathway that is responsible for most of the ATP formed during cellular respiration is
   a. glycolysis.
   b. the Krebs cycle.
   c. the transition reaction.
   d. the electron transport system.

____ 23. The pathway that begins with glucose and ends with pyruvate is _____, and it occurs in the _____.
   a. glycolysis; cytoplasm
   b. the Krebs cycle; mitochondria
   c. the transition reaction; cytoplasm
   d. glycolysis; cristae of the mitochondria

____ 24. The electron transport system is located where?
   a. within the matrix of the mitochondria
   b. on the rough endoplasmic reticulum
   c. on the ribosomes
   d. on the cristae of the mitochondria

____ 25. The final acceptor for hydrogen in aerobic cellular respiration is
   a. ATP.
   b. NAD.
   c. FAD.
   d. oxygen.

THOUGHT QUESTIONS

Answer in complete sentences.

26. What happens when no oxygen is present in the cell at the end of glycolysis?

27. If a cell is placed into a hypertonic environment, what is the result?

Test Results: _____ number correct ÷ 27 = _____ × 100 = _____ %

ANSWER KEY

STUDY QUESTIONS

1. a. surface area **b.** volume **2. a.** simple compound **b.** scanning **c.** transmission **d.** either electron microscope **3. a.** glycoprotein **b.** carbohydrate chain **c.** glycolipid **d.** protein molecule **e.** phospholipid bilayer **f.** cytoskeleton filaments **g.** cholesterol **4.** regulates **5. a.** polar **b.** hydrophobic **6. a.** Diffusion **b.** Osmosis **7. a.** isotonic **b.** hypotonic **8. a.** Facilitated **b.** Active **9. a.** endocytosis **b.** Exocytosis **10. a.** centriole: formation of basal bodies **b.** Golgi apparatus: processing, packaging, and distributing molecules **c.** microtubule: portion of cytoskeleton **d.** mitochondrion: cellular respiration **e.** rough ER: protein synthesis **f.** ribosome: protein synthesis **g.** vacuole or vesicle: storage of substances **h.** lysosome: intracellular digestion **i.** smooth ER: various; lipid synthesis in some cells **j.** nucleus: storage of genetic information **k.** nucleolus: ribosomal formation **11. a.** nuclear envelope **b.** pores **c.** nucleolus **d.** ribosomes **12. a.** rough **b.** Golgi apparatus **13. a.** microtubules **b.** actin filaments **c.** basal bodies **d.** cilia **e.** flagella **14. a.** metabolic pathway **b.** enzyme **15. a.** ES **b.** enzyme **c.** substrate **d.** product **16. a.** coenzymes **b.** NAD **17. a.** cristae **b.** matrix **c.** matrix **d.** cristae **e.** in the cytoplasm outside the mitochondrion **18. a.** pyruvate **b.** a C_2 acetyl group + CO_2 **c.** Krebs cycle **d.** electron transport system **19.** electron transport system **20. a.** lactic acid **b.** fermentation

DEFINITIONS WORDSEARCH

a. centriole **b.** chromosome **c.** cilium **d.** flagellum **e.** organelle **f.** nucleolus **g.** microtubule **h.** coenzyme **i.** cytoskeleton **j.** cytoplasm

```
                                              M
  E M O S O M O R H C                         I
                  Y                           C
                  T                           R
            C Y T O S K E L E T O             N
      F           P             S             T
      L           L       E U                 U
      A           A       L                   B
      G           S       O O                 U
      C O E N Z Y M E     I                   L
      L           L       R                   E
      L           C       T
      U           U       N
  C I L I U M           E L L E N A G R O
                          C
```

CHAPTER TEST

1. c **2.** d **3.** c **4.** b **5.** d **6.** a **7.** b **8.** d **9.** c **10.** a **11.** d **12.** b **13.** d **14.** e **15.** c **16.** a **17.** b **18.** d **19.** f **20.** b **21.** a **22.** d **23.** a **24.** d **25.** d **26.** When no oxygen is present, the pyruvate cannot enter the transition reaction and be transported into the mitochondria to engage in the Krebs cycle. Instead, the cell turns to fermentation pathways to release a small quantity of energy. The result is the buildup of lactic acid within the cell. **27.** A cell placed in a hypertonic environment will shrink as it loses water to its surroundings.

3

INTRODUCTION TO HEMCOSTASIS

STUDY TIPS

In the chapters that follow, the various organ systems are discussed. Having a good understanding of tissues will help you to appreciate the complexity of organs and organ systems. Homeostasis is a recurring theme throughout the remaining chapters. Many homeostatic mechanisms operate by negative feedback, much as a thermostat regulates the temperature of a room. Keep this in mind as an example of how homeostasis works as you study upcoming chapters.

Biological terminology is often confusing, especially to a student new to the discipline. In this chapter, you will note the use of the term *fibers* as reinforcing protein strands within connective tissue (p. 64). Fibers also refer to muscle cells. Be aware that it is easy to confuse the two. Another example of a term having more than one definition is the use of *membranes* to indicate structures within or surrounding the cell, and the use of membranes in this chapter in reference to body linings (p. 69). Be sure to determine the context in which the term **membrane** is used.

STUDY QUESTIONS

Study the text section by section as you answer the questions that follow.

3.1 TYPES OF TISSUES (PAGES 62–68)

- Describe the features and functions of epithelial tissues.
- Explain how junctions between cells aid communication between them.
- Discuss the different types of connective tissues and their functions.
- Describe the types of muscle tissue and how they operate.
- State the characteristics and functions of nervous tissue.

1. Draw a diagram of squamous epithelium.
   a.

   Name one place in the human body where squamous epithelium can be found. b. _Skin_
   What is the function of this tissue? c. _To protect any physical damage_

   When this tissue is found in layers, it is called d. _epithelial_ squamous epithelium.

2. Epithelium is classified by the a._____ of its cells. b._____ epithelium has cube-shaped cells, while c._____ epithelium has elongated, cylindrical cells.

3. The windpipe is lined by pseudostratified ciliated columnar epithelium. Describe this tissue.
   _____
   _____
   _____

4. a._Membranes_ junctions allow for materials and information to be exchanged between cells. b._____ junctions are a means of reinforcement in which cytoskeleton filaments are woven together. c._____ junctions are similar to spot welds as they bind cells firmly together.

5. Loose fibrous connective tissue has cells called a._____, plus fibers made from b._____ and c._____. It binds the skin to underlying organs.

6. Tendons and ligaments are made up of a._____ tissue. Tendons join b._____ to bone and ligaments join c._____ to bone.

7. a._____ cartilage can be found at the ends of bones and makes up the fetal skeleton. In it, cells lie within b._____, surrounded by a gel-like c._____.

8. The type of cartilage found in the ear, called a._____ cartilage, has many b._____ fibers to add flexibility. c._____, found in the intervertebral disks, aids in cushioning against jolts.

9. Draw a sketch of bone tissue. Label the lacunae, the matrix, and the osteons (Haversian systems).

10. Blood is considered a connective tissue because its cells are separated by a fluid matrix called a._____. b._____ blood cells carry oxygen throughout the body in association with molecules of hemoglobin. c._____ blood cells are responsible for immunity, and cell fragments called d._____ aid in blood clotting.

11. Complete this table.

	Fiber Appearance	Location	Control
Skeletal	a.		
Smooth	b.		
Cardiac	c.		

12. The brain and spinal cord are made up of cells called a._____. Outside the central nervous system, connective tissue binds the long fibers of these cells to form b._____.
    The function of a neuron is to c._____.
    The other type of cell in nervous tissue is d._____ cells.
    This type of cell provides e._____ to neurons and keeps tissue free of debris.
    Neuroglial cells outnumber neurons f._____ to one and take up more than g._____ the volume of the brain.

3.2 Body Cavities and Body Membranes (page 69)

- Describe the membranes of the body cavities.

13. Place a T in front of all organs that are in the thoracic cavity, an A in front of all organs that are in the abdominal cavity, and a P for those that are in the pelvic cavity.

   a. __A__ small intestine
   b. __P__ ovaries
   c. __P__ bladder
   d. __T__ heart
   e. __T__ lungs
   f. __A__ stomach
   g. __A__ liver
   h. __A__ kidneys

14. A large, ventral cavity called a(n) a._____ can be seen during embryological development. It later develops into the b._____ cavity of the chest and the c._____ cavity of the lower abdomen.

15. The a._____ are connective tissue membranes protecting the brain and spinal cord. b._____ lines the gut and secretes mucus. c._____ membrane lines joint cavities and secretes lubricating d._____.

16. Pleural membranes and peritoneum are examples of a._____ membranes. These secrete b._____ to keep the membranes moist and prevent sticking.

3.3 Organ Systems (page 70)

- Give examples of the various organ systems of the body and their general functions.

17. The a. _Digestion_ system is responsible for reducing food to molecules that can be used by the cells of the body. The b. _Muscular_ system allows us to move from one place to another and generates heat. The c. _Nervous_ system consists of the brain, spinal cord, nerves, and receptors. The d. _Connective_ system sends out hormones to regulate bodily functions.

18. The *Working Together* diagram on p. 71 of your textbook illustrates the ways that the systems of the body contribute to homeostasis. In similar diagrams in the following chapters, how the organ systems work together to help each other maintain homeostasis will be shown. Match the organ systems with the correct descriptions.

   a. Breakdown and absorption of food materials
   b. Gaseous exchange between external environment and blood
   c. Regulation of all body activities; learning and memory
   d. Body movement; production of body heat
   e. External support and protection of body
   f. Secretion of hormones for chemical regulation
   g. Production of sperm; transfer of sperm to female reproductive system
   h. Transport of nutrients to body cells; removal of wastes from cells
   i. Immunity; absorption of fats; drainage of tissue fluid
   j. Maintenance of volume and chemical composition of blood
   k. Internal support and protection; body movement; production of blood cells

   __e__ integumentary system
   __b__ respiratory system
   __i__ lymphatic system
   __a__ digestive system
   __k__ skeletal system
   __c__ nervous system
   __j__ urinary system
   __g__ reproductive system
   __f__ endocrine system
   __d__ muscular system
   __h__ cardiovascular

3.4 Skin as an Organ System (pages 72–73)

- List the three regions of the skin, their functions, and the structures located within them.
- Describe how body temperature is regulated.

19. A(n) a. _Organ_ is made up of two or more tissues, functioning together. The skin is an example. It has an outer layer called the b. _epidermis_. The cells of this layer become waterproof once they are filled with the protein c. _keratin_.

20. An accessory organ of the skin, the a. _____, forms hair shafts. b. _____ muscles cause goose bumps to appear. c. _____ glands provide sebum to moisturize hairs.

21. Label the following illustration of the skin, using the alphabetized list of terms.
    - adipose tissue
    - arrector pili muscle
    - blood vessels K
    - connective tissue e
    - dermis b
    - epidermis a
    - hair root h
    - hair shaft d
    - nerve f
    - oil gland g
    - receptors
    - subcutaneous layer c
    - sweat gland l

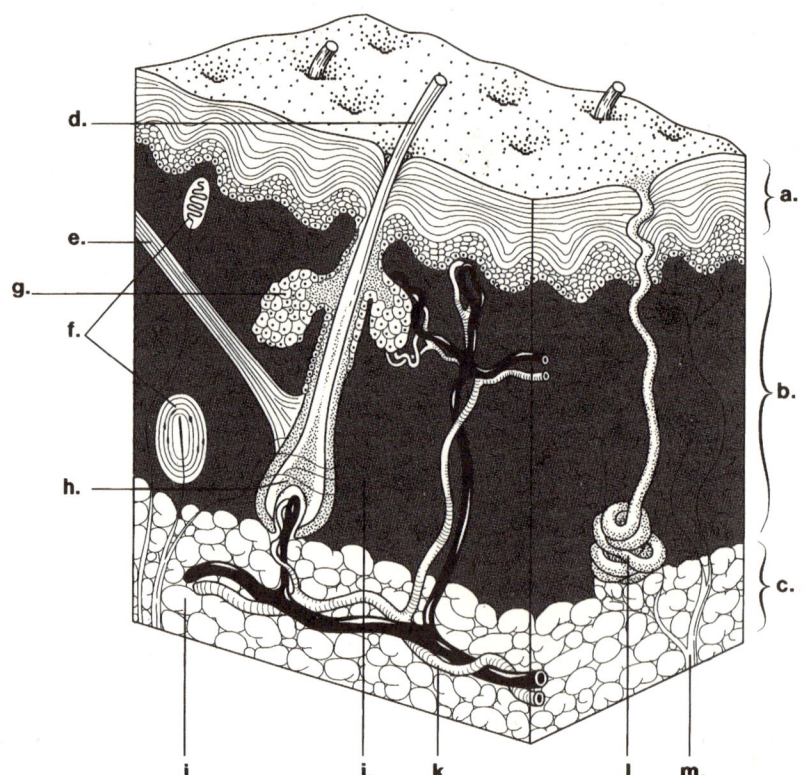

22. What events raise body temperature? a. _____

    What events lower body temperature? b. _____

3.5 HOMEOSTASIS (PAGES 76–77)

- Define homeostasis, tell how it is controlled, and explain its importance.
- Understand how the nervous and endocrine systems coordinate to maintain homeostasis.
- Use the maintenance of room temperature as an example of negative feedback control.

23. What is homeostasis? a. *the constantance of the internal envi iregardless of the external change*
    What is the internal environment? b. _____
    Give an example of homeostasis. c. *body temperature of 37°C*

24. How does each of the following systems contribute to homeostasis?
    digestive system a. _____
    respiratory system b. _____
    urinary system (i.e., the kidneys) c. _____

25. The two systems of the body that control homeostasis are a. *Reception*
    and b. *Regulatory centre*.

26. Use the following terms to complete this diagram of a negative feedback control: *effector, sensory receptor, regulatory center,* and *response*.

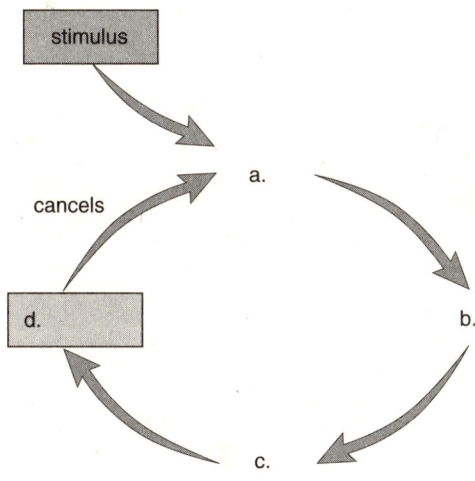

27. Give an example of negative feedback control in the body. _____
_____
_____

DEFINITIONS WORDSEARCH

Review key terms by completing this wordsearch, using the following alphabetized list of terms:

```
E H J Y R D E R M I S Y T X
P T W F E D P P O P U L I C
I W E D I L I G A M E N T L
T A S R T W D I O P T L N J
H M I B O N E G U B V E R O
E J T L B C R N O R U E N L
L C O E L O M K R L D K I D
I O F E T I I H U G N U F O
A M S E F Y S T R I A T E D
L P M I N D L E S H L I F B
C A R T I L A G E O G I F T
D C P T J U N E T F D I L T
C T E N D O N J U N P T I F
```

bone
cartilage
coelom
compact
dermis
epidermis
epithelial
gland
ligament
neuron
striated
tendon

a. __epithelial__ Type of tissue that covers the body and lines its cavities.
b. __dermis__ Inner layer of skin that lies beneath the epidermis.
c. _____ Fibrous connective tissue joining bone to bone.
d. _____ Rigid connective tissue containing mineral salts.
e. __epidermis__ Outer layer of skin, protective in nature.
f. _____ Specialized epithelial cell that secretes substances.
g. _____ Having alternating light and dark bands.
h. _____ Connective tissue found on ends of bones or in the ear.
i. _____ Fibrous connective tissue connecting muscle to bone.
j. _____ Type of bone with densely packed osteons.
k. _____ Embryonic body cavity that becomes the thoracic and abdominal cavities.
l. _____ A nerve cell.

CHAPTER TEST

OBJECTIVE TEST

Do not refer to the text when taking this test.

__A__ 1. Cells of a similar structure working together constitute a(n)
   a. tissue.
   b. organ.
   c. organism.
   d. organ system.

__d__ 2. Epithelial tissues do which of the following?
   a. secrete
   b. line body cavities
   c. protect
   d. all of these

__d__ 3. Which of these pairs is mismatched?
   a. fat–subcutaneous layer
   b. receptors–dermis
   c. keratinization–epidermis
   d. nerves and blood vessels–epidermis

__c__ 4. If epithelial tissue is made up of many layers of cells, what would be true?
   a. It is called simple.
   b. It is called stratified.
   c. It is called striated.
   d. It is pseudostriated.

28

__b__ 5. Which type of epithelial tissue is composed of flattened cells?
   a. glandular
   b. squamous
   c. columnar
   d. cuboidal

__c__ 6. What type of tissue binds skin to underlying organs?
   a. fibrous connective tissue
   b. nervous tissue
   c. muscular tissue
   d. loose connective tissue

_____ 7. Which type of cell junction is characterized by interwoven cytoskeleton filaments?
   a. tight junction
   b. gap junction
   c. adhesion junction
   d. intercalated disks

_____ 8. Which of the following tissues has cells residing in lacunae?
   a. adipose tissue
   b. fibrous connective tissue
   c. hyaline cartilage
   d. loose connective tissue

_____ 9. Osteocytes are residents of _____ tissue.
   a. cartilage
   b. bone
   c. muscle
   d. pseudostratified columnar epithelium

For questions 10–13, match the following blood components to the appropriate statement.
   a. plasma   b. red blood cells
   c. white blood cells   d. platelets

__d__ 10. contribute to blood clotting process
__b__ 11. carry oxygen to cells
__a__ 12. fluid matrix component of the connective tissue called blood
__c__ 13. are responsible for immunity

For statements 14–17, match the following components of skin to the appropriate statement.
   a. epidermis   b. arrector pili   c. melanocytes
   d. oil glands

__d__ 14. when these become blocked, a blackhead forms
__c__ 15. give rise to skin pigment
__b__ 16. muscle attached to a hair follicle
__a__ 17. keratinized layer of skin

__d__ 18. Cells in the nervous tissue that support and protect neurons are called
   a. neurons.
   b. fibers.
   c. axons.
   d. neuroglial cells.

__c__ 19. Cardiac muscle fibers are characterized by which traits?
   a. striated; intercalated disks
   b. smooth; single nucleus
   c. striated; multiple nuclei; intercalated disks
   d. smooth; tapered; multinucleate multiple nuclei

__c__ 20. Which of these is located in the thoracic cavity?
   a. small intestine
   b. urinary bladder
   c. lungs
   d. kidneys

_____ 21. The thoracic cavity is lined with
   a. mucous membrane.
   b. serous membrane.
   c. cutaneous membrane.
   d. synovial membrane.

_____ 22. Which type of membrane lines joint cavities?
   a. mucous membrane
   b. serous membrane
   c. cutaneous membrane
   d. synovial membrane

__c__ 23. Which of these is an example of homeostasis?
   a. Muscle tissue is specialized to contract.
   b. The skin sunburns.
   c. Normal body temperature almost always stays at 37 °C.
   d. All of these

_____ 24. In a negative feedback control system,
   a. homeostasis is impossible.
   b. there is a constancy of the internal environment.
   c. there is a wide fluctuation occurring continuously.
   d. none of the above is true.

__c__ 25. When body temperature rises, sweat glands become
   a. active, and blood vessels constrict.
   b. inactive, and blood vessels dilate.
   c. active, and blood vessels dilate.
   d. inactive, and blood vessels constrict.

THOUGHT QUESTIONS

Answer in complete sentences.

26. How do the nervous system and endocrine system interact to maintain the homeostasis of the body?

27. How do the blood vessels, the digestive tract, the lungs, and the kineys work together to maintain homeostasis?

Test Results: _____ number correct ÷ 27 = _____ × 100 = _____%

Answer Key

Study Questions

1. a. see Figure 3.2, in text **b.** air sacs of the lungs **c.** absorption **d.** stratified **2. a.** shape **b.** Cuboidal **c.** columnar **3.** This type of tissue appears to be layered but is not, and cells have small, hairlike projections called cilia. **4. a.** Gap **b.** Adhesion **c.** Tight **5. a.** fibroblasts **b.** collagen **c.** elastin **6. a.** dense fibrous connective **b.** muscle **c.** bone **7. a.** Hyaline **b.** lacunae **c.** matrix **8. a.** elastic **b.** elastic **c.** Fibrocartilage **9.** see Figure 3.4, in text **10. a.** plasma **b.** Red **c.** White **d.** platelets **11. a.** striated; skeleton; voluntary **b.** spindle-shaped; internal organs; involuntary **c.** striated; heart; involuntary **12. a.** neurons **b.** nerves **c.** conduct nerve impulses **d.** neuroglial cells **e.** protection **f.** nine **g.** one-half **13. a.** A **b.** P **c.** P **d.** T **e.** T **f.** A **g.** A **h.** A **14. a.** coelom **b.** thoracic **c.** abdominal **15. a.** meninges **b.** Mucous membrane **c.** Synovial **d.** synovial fluid **16. a.** serous **b.** watery serous fluid **17. a.** digestive **b.** muscular **c.** nervous **d.** endocrine **18.** e, b, i, a, k, c, j, g, f, d, h **19. a.** organ **b.** epidermis **c.** keratin **20. a.** hair follicle **b.** Arrector pili **c.** Oil **21. a.** epidermis **b.** dermis **c.** subcutaneous layer **d.** hair shaft **e.** arrector pili muscle **f.** receptors **g.** oil gland **h.** hair root **i.** adipose tissue **j.** connective tissue **k.** blood vessels **l.** sweat gland **m.** nerve **22. a.** Blood vessels constrict, sweat glands are inactive, hair stands on end, and shivering may occur. **b.** Blood vessels dilate, sweat glands are active, and hair lies next to skin. **23. a.** relative constancy of the internal environment **b.** tissue fluid **c.** body temperature remains around 37° C **24. a.** provides nutrient molecules **b.** removes carbon dioxide and adds oxygen to the blood **c.** eliminates wastes and salts **25. a.** nervous **b.** endocrine **26. a.** sensory receptor **b.** regulatory center **c.** effector **d.** response **27.** Sensory receptors in aortic and carotid sinuses communicate with a regulatory center in the brain when there is a rise in blood pressure. This center sends inhibitory nerve impulses to the arteries, and they relax. When the blood pressure falls, the sensory receptors no longer communicate with the brain center.

Definitions Wordsearch

```
E           D E R M I S
P           P
I           L I G A M E N T
T           D
H     B O N E
E           R N O R U E N
L C O E L O M       D
I O         I       N
A M         S T R I A T E D
L P                 L
C A R T I L A G E   G
C
T E N D O N
```

a. epithelial **b.** dermis **c.** ligament **d.** bone **e.** epidermis **f.** gland **g.** striated **h.** cartilage **i.** tendon **j.** compact **k.** coelom **l.** neuron

Chapter Test

1. a **2.** d **3.** d **4.** b **5.** b **6.** d **7.** c **8.** c **9.** b **10.** d **11.** b **12.** a **13.** c **14.** d **15.** c **16.** b **17.** a **18.** d **19.** a **20.** c **21.** b **22.** d **23.** c **24.** b **25.** c **26.** The nervous system is the ultimate source of control over homeostasis—it registers changes and decides what action to take. Nervous impulses can be sent to trigger an appropriate response to maintain homeostasis. The endocrine system releases hormones into the bloodstream when needed to maintain homeostasis. Hormones last longer, but nervous impulses can travel faster. The nervous system controls the endocrine system. **27.** The digestive tract adds nutrients and the lungs add oxygen to the blood. The blood vessels bring nutrients and oxygen to tissue fluid and take away waste molecules, including carbon dioxide. The kidneys excrete metabolic wastes and the lungs excrete carbon dioxide.

PART 2 MAINTENANCE OF THE HUMAN BODY

4

DIGESTIVE SYSTEM AND NUTRITION

STUDY TIPS

Begin your study of this chapter by surveying the organs of the digestive system (pp. 82–89), and discover how each is uniquely adapted to carrying out a function that contributes to the overall process of digestion. Next, study the contributions of accessory organs (pp. 90–91) and digestive enzymes (pp. 92–93) that complete the chemical breakdown of nutrients.

A knowledge of nutrition is essential for good health and disease prevention. More and more information comes to light daily about what we should and should not eat.

Following the basic nutrition guidelines set forth in this chapter (pp. 95–105) will contribute to a healthy lifestyle.

Keep in mind as you study this chapter that the purpose of the digestive system is to break down complex foods into nutrients that can be used by your body's *cells*. You learned in chapter 2 that your body functions at the level of the cell. The digestive system is a series of organs that provide nutrients for cells.

STUDY QUESTIONS

Study the text section by section as you answer the questions that follow.

4.1 THE DIGESTIVE SYSTEM (PAGES 82–89)

- Trace the path of food from mouth to anus during digestion.
- Describe the features of the mouth that prepare food for swallowing.
- Explain how the structure of the pharynx ensures the passage of food into the esophagus.
- Discuss the features of the esophagus and how peristalsis moves food along this structure.
- Explain how the structure of the stomach is adapted to high acidity.
- Describe the structure and function of the four layers of the digestive tract.
- Relate the function of the small intestine to its unique structure.
- Explain how microvilli enhance absorption.
- Discuss how digestive secretions are under hormonal control, and list the hormones involved.
- Describe the sections of the large intestine, and know its general function.

Questions 1–7 pertain to the following diagram.

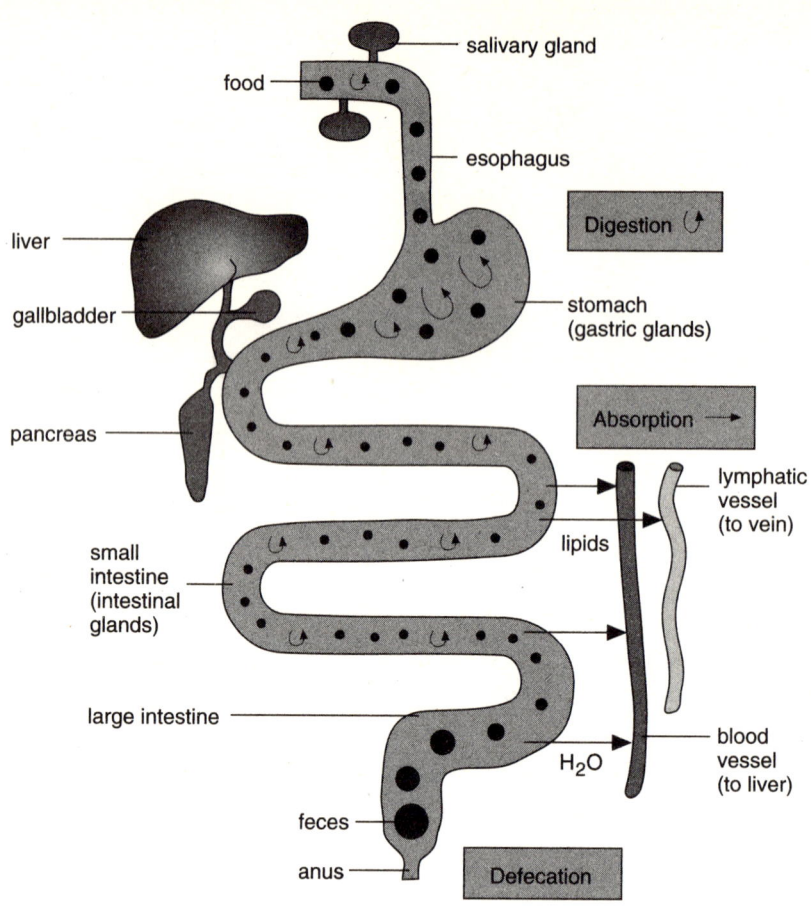

1. Food is received by the a._mouth_____, which possesses b._teeth_____ to grind and break up particles. c._Saliva____ glands secrete d._saliva_____ that moistens food and binds it together for swallowing. The taste of food is detected by e._____ located on the f._nose_____.
2. At the back of the mouth, the muscular a._pharynx_____ is involved in swallowing food. The epiglottis folds over the glottis during the swallowing b._trachea_____.
3. The esophagus propels food toward the a._____, using muscular contractions called b._____. At the upper end of the esophagus, a muscular ring called a(n) c._lower_____ regulates the entrance of food to the esophagus.
4. The stomach is lined with a._mucos_____ that secretes protective b._____. Otherwise, the lining would be injured by the strong c._____ acid secreted by d._____ glands. Gastric juice also contains e._____ that breaks down proteins.
5. The layer of the digestive tract that houses blood vessels is the a._____. The b._____ layer contains two layers of smooth muscle. The outside layer, c._____, protects the digestive tube.
6. The first section of the small intestine is the a._____. It receives secretions from the b._____ and c._____ and also receives food known as d._____ from the stomach.
7. The mucosa of the small intestine is folded into a._____, which, in turn, have projections from individual cells called b._____.
Proteins in the microvilli transport nutrients such as c._____ directly into the blood stream, and fats must be reconstructed so they can travel into vessels of the lymphatic system called d._____.

32

8. State the purpose of microvilli. _____
   _____

9. A number of hormones control the secretions of digestive juices. a. _____ is secreted in response to protein in foods and enhances gastric gland output, while b. _____ inhibits gastric secretion. c. _____ is secreted in response to acidic chyme. When fats are present in chyme, d. _____ triggers the release of bile from the gallbladder.

10. The large intestine functions to store and compact a. _____. Unusual outgrowths of the lining of the colon, called b. _____, can be either benign or cancerous. Colon cancer incidence increases for people who do not have enough c. _____ in their diets.

4.2 Three Accessory Organs (pages 90–91)

- Explain how the secretions from the pancreas aid digestion.
- List several functions of the liver, and describe the role of the gallbladder in digestion.
- Discuss the major diseases of the liver and their causes.

11. Pancreatic juice contains a mix of a. _bicarbonate_ / _hydrochloric acid_ solution to neutralize stomach acid, and digestive b. _food_ to further break down food.

12. The liver produces a greenish substance called a. _bile_, which is stored and concentrated by the b. _gallbladder_. The liver has been called the c. _____ to the blood because it detoxifies substances entering the blood from the d. _____. Other functions of the liver include (list three)
    e. _Produce bile_
    f. _make plasma protein_
    g. _Remove harmful substance e.g. alcohol_

13. a. _____ is an inflammatory disease of the liver caused by a viral infection, while
    b. _____ is damage caused by chronic alcohol abuse.

4.3 Digestive Enzymes (pages 92–93)

- Name the major digestive enzymes and the types of nutrients they decompose, and state in which organs they are produced. _types - protein   maltose - maltase_

14. a. _____ is a fat-digesting enzyme produced by the accessory organ, the b. _____. Starch digestion begins in the c. _____ with salivary d. _____ and continues in the e. _____ with pancreatic f. _____.

15. Protein digestion begins in the a. _____ where the enzyme b. _____ is produced, and continues in the small intestine with the aid of the enzyme c. _____ from the pancreas.

4.4 HOMEOSTASIS (PAGE 95)

- Discuss how the digestive system works with other systems of the body to maintain homeostasis.

Use the following information to help you interpret *Working Together* diagrams.

Each *Working Together* diagram shows how one particular body system interacts with other body systems. In the example that follows, the **digestive system** is the main system under discussion. How it benefits the muscular system and receives benefits from the muscular system is also described in the figure on page 94 of your textbook.

The top portion tells how the digestive system benefits the muscular system.

The bottom portion tells how the muscular system benefits the digestive system.

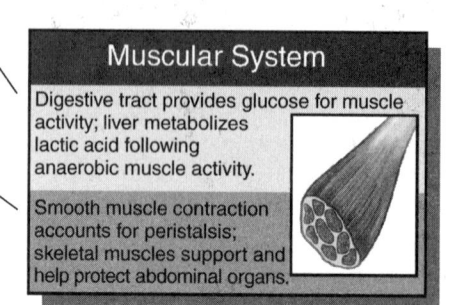

Muscular System

Digestive tract provides glucose for muscle activity; liver metabolizes lactic acid following anaerobic muscle activity.

Smooth muscle contraction accounts for peristalsis; skeletal muscles support and help protect abdominal organs.

16. The *Working Together* diagram in your textbook illustrates the ways in which the digestive system benefits other organ systems of the body. Match the organ systems with the correct descriptions.
    a. *Digestive tract provides nutrients for plasma protein formation and blood cell formation; liver detoxifies blood, makes plasma proteins, destroys old red blood cells.*
    b. *Liver synthesizes urea; digestive tract excretes bile pigments from liver and provides nutrients.*
    c. *Digestive tract provides nutrients needed by the skin.*
    d. *Digestive tract provides nutrients for growth, maintenance, and repair of neurons and neuroglial cells.*
    e. *Digestive tract provides glucose for muscle activity; liver metabolizes lactic acid following anaerobic muscle activity.*
    f. *Breathing is possible through the mouth because digestive tract and respiratory tract share the pharynx.*
    g. *Stomach and small intestine produce hormones.*
    h. *Digestive tract provides nutrients for lymphoid organs; stomach acidity prevents pathogen invasion of the body.*
    i. *Digestive tract provides nutrients for growth and repair of organs and for development of fetus.*
    j. *Digestive tract provides Ca^{++} and other nutrients for bone growth and repair.*

    _____ integumentary system
    _____ respiratory system
    _____ lymphatic system
    _____ circulatory system
    _____ skeletal system
    _____ nervous system
    _____ urinary system
    _____ reproductive system
    _____ endocrine system
    _____ muscular system

4.5 NUTRITION (PAGES 95–105)

- Describe how the different classes of nutrients enter into general circulation within the body.
- Discuss proper nutrition and how carbohydrates, protein, and fat should be proportioned in the diet.
- Discuss the vitamin and mineral requirements in the diet.

17. Fill in the blanks of the following food guide pyramid. Include the following groups, listed in alphabetical order, and indicate the recommended number of daily servings from each group:

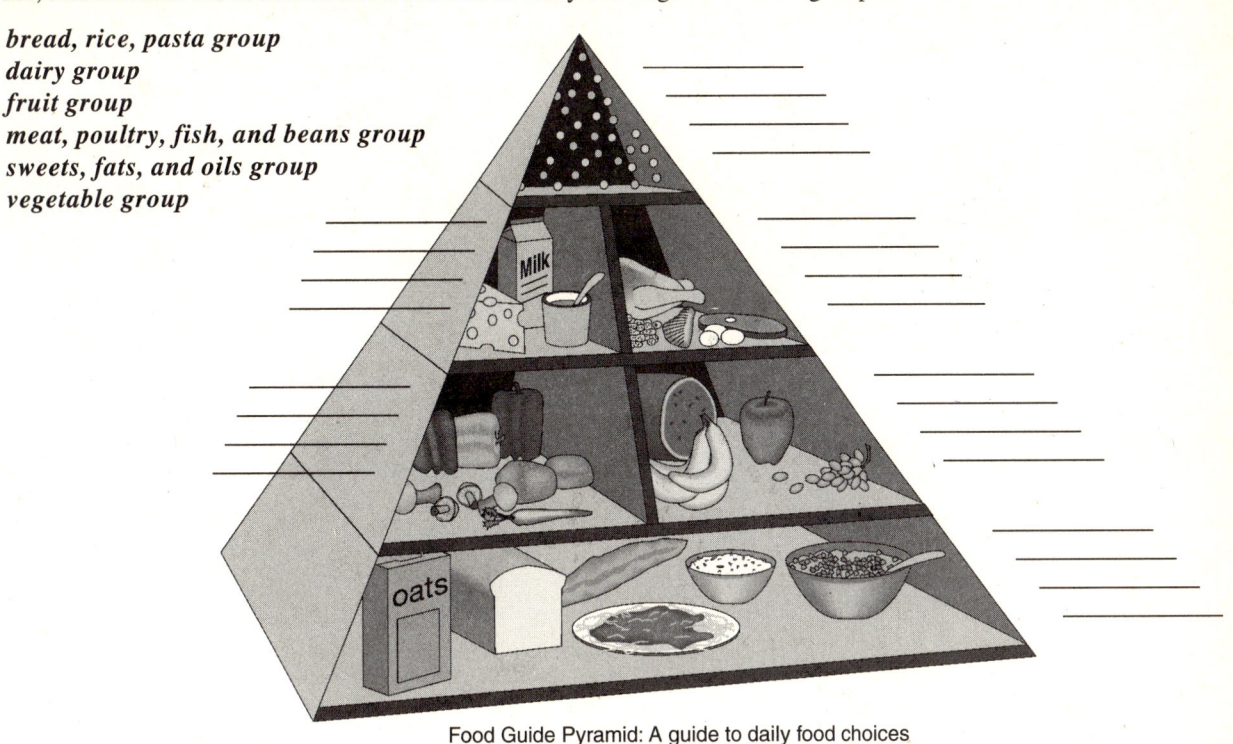

   bread, rice, pasta group
   dairy group
   fruit group
   meat, poultry, fish, and beans group
   sweets, fats, and oils group
   vegetable group

   Food Guide Pyramid: A guide to daily food choices

18. Of the macronutrients, a._____ are the primary source of energy in the diet, while b._____ supply building materials. c._____ are stored to supply energy at a later time.

19. In the small intestine, a._____ and b._____ pass into the wall of the small intestine and from there, into c._____. Fats travel a different route, into d._____, after being reconstructed in intestinal cells.

20. It is recommended that carbohydrates comprise a._____% of calories in the diet, with the bulk of those being b._____ carbohydrates. Fats should be less than c._____% of total calories for the day, especially keeping d._____ fats to a minimum.

21. Many vitamins function as a._____ in various metabolic pathways. Vitamin b._____ is important as a visual pigment, and vitamin c._____ is turned into a compound that enhances calcium absorption. Vitamins d._____ are antioxidants.

22. Too much of the mineral a._____ can lead to hypertension. The mineral b._____ is a major component of bones and teeth.

Eating Disorders (Pages 104–105)

23. Assign an *O* for obesity, *B* for bulimia nervosa, or an *A* for anorexia nervosa to these descriptions.
    _____ a. Body weight is OK, but weight is regulated by purging after eating.
    _____ b. Body weight is too low, and person may binge eat and purge.
    _____ c. Body weight is 20% or more above appropriate weight for height.
    _____ d. Exercise is most often minimal.
    _____ e. Often accompanied by excessive exercise.

DEFINITIONS CROSSWORD

Review key terms by completing this crossword puzzle using the following alphabetized list of terms:

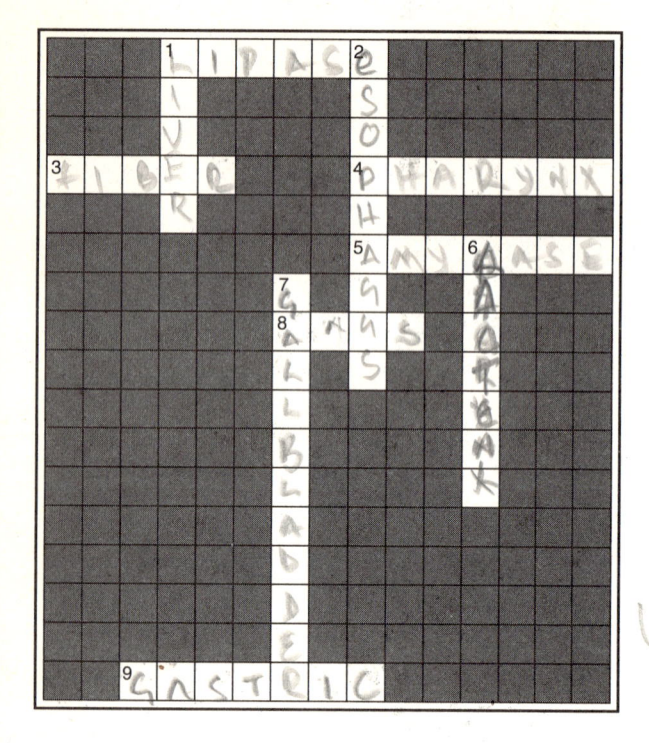

amylase
anus
esophagus
fiber
gallbladder
gastric
lacteal
lipase
liver
pharynx

Across
1 Fat-digesting enzyme secreted by the pancreas.
3 Indigestible plant material that lowers risk of colon cancer.
4 Muscular passageway at the back of the mouth where swallowing occurs.
5 Starch-digesting enzyme.
8 Final canal of the large intestine.
9 Type of gland found in the stomach.

Down
1 Large, multifunctioned organ in the abdominal cavity, secretes bile.
2 Muscular tube leading from the pharynx to the stomach.
6 Lymphatic vessel inside a villus of the small intestine.
7 Muscular sac that stores bile.

CHAPTER TEST

OBJECTIVE TEST

Do not refer to the text when taking this test.
    For questions 1–6, match the following digestive system structures to the appropriate statements.
    a. pharynx   b. soft palate   c. esophagus
    d. stomach   e. small intestine   f. large intestine

___ 1. blocks food from entering nasal passages
___ 2. conducts food from throat to stomach
___ 3. common passageway for food and air
___ 4. longest portion of the digestive tract
___ 5. contains cecum, colon, rectum, and anus
___ 6. has pyloric sphincter on one end
___ 7. The layer of the digestive tube that keeps it from sticking to surrounding internal organs is the
   a. serous layer.
   b. submucosa.
   c. mucosa.
   d. muscularis.

For questions 8–12, match these organs to the appropriate function.

    a. large intestine b. stomach c. liver
    d. pancreas e. gallbladder

___d___ 8. secretes digestive enzymes and bicarbonate

___e___ 9. releases bile into the duodenum

___b___ 10. absorbs water and alcohol, initiates protein breakdown

___c___ 11. detoxifies blood

___a___ 12. absorbs water and houses bacteria

___b___ 13. Cholecystokinin is a(n) _____ secreted in response to _____ in food.
    a. enzyme, protein
    b. hormone, fats
    c. lipid, carbohydrates
    d. hormone, carbohydrates

___d___ 14. Which enzyme digests starch?
    a. lipase
    b. carboxypeptidase
    c. trypsin
    d. amylase

___b___ 15. HCl
    a. is an enzyme.
    b. creates the acid environment necessary for pepsin to work.
    c. is found in the intestines.
    d. all of these.

___d___ 16. Two enzymes involved in the digestion of protein are
    a. salivary amylase and lipase.
    b. trypsin and hydrochloric acid.
    c. pancreatic amylase and trypsin.
    d. pepsin and trypsin.

___b___ 17. What is absorbed into the lacteals?
    a. proteins
    b. fats
    c. carbohydrates
    d. water and amino acids

___d___ 18. Villi
    a. are found in the small intestine.
    b. increase the absorptive surface area.
    c. contain capillaries.
    d. All of these are correct.

___d___ 19. The enzymes for digestion are referred to as hydrolytic because they require
    a. hydrogen.
    b. HCl.
    c. energy.
    d. water.

___a___ 20. What type of food has the highest fat content?
    a. meat
    b. bread
    c. potatoes
    d. fruits

___b___ 21. What is the best way to ensure that you are obtaining plenty of vitamins and minerals in your diet?
    a. Take mega-multiple vitamin supplements.
    b. Be sure you eat five fruit and vegetable servings daily.
    c. Eat lots of red meat.
    d. Eat white bread.

___c___ 22. Which macronutrient supplies quick energy and should comprise the bulk of the diet?
    a. milk and cheese
    b. lipids
    c. carbohydrates
    d. proteins

___a___ 23. An eating disorder characterized by binging on food, then inducing vomiting or other purging is
    a. bulimia.
    b. anorexia nervosa.
    c. obesity.
    d. cirrhosis.

___d___ 24. Which type of fat is best to consume in quantity?
    a. saturated fats
    b. monounsaturated fats
    c. polyunsaturated fats
    d. None of these should be consumed in quantity.

___d___ 25. Which of the following are considered antioxidants?
    a. calcium and sodium
    b. B vitamins and selenium
    c. vitamin E and iron
    d. vitamins C, E, and A

THOUGHT QUESTIONS

Answer in complete sentences.

26. Why is digestion a necessary process for humans?

27. What are LDL and HDL cholesterol?

28. How do the digestive and reproductive systems interact?

Test Results: _____ number correct ÷ 28 = _____ × 100 = _____%

ANSWER KEY

STUDY QUESTIONS

1. a. mouth **b.** teeth **c.** Salivary **d.** saliva **e.** taste buds **f.** tongue **2. a.** pharynx **b.** reflex **3. a.** stomach **b.** peristalsis **c.** sphincter (constrictor) **4. a.** mucosa **b.** mucus **c.** hydrochloric **d.** gastric **e.** pepsin **5. a.** submucosa **b.** muscularis **c.** serosa **6. a.** duodenum **b.** pancreas **c.** gallbladder **d.** chyme **7. a.** villi **b.** microvilli **c.** glucose and amino acids **d.** lacteals **8.** To increase the surface area available for absorption in the small intestine. **9. a.** gastrin **b.** gastric inhibitory peptide **c.** Secretin **d.** cholecystokinin **10. a.** feces **b.** polyps **c.** fiber **11. a.** bicarbonate **b.** enzymes **12. a.** bile **b.** gallbladder **c.** gatekeeper to blood **d.** small intestine **e.** storing iron and fat-soluble vitamins **f.** storing glycogen **g.** making blood plasma proteins **13. a.** Hepatitis **b.** cirrhosis **14. a.** Lipase **b.** pancreas **c.** mouth **d.** amylase **e.** small intestine **f.** amylase **15. a.** stomach **b.** pepsin **c.** trypsin **16.** c, f, h, a, j, d, b, i, g, e **17.** see Figure 4.13, in text **18. a.** carbohydrates **b.** proteins **c.** Lipids **19. a.** carbohydrates **b.** proteins **c.** blood **d.** lacteals **20. a.** 58 **b.** complex **c.** 30 **d.** saturated **21. a.** coenzymes **b.** A **c.** D **d.** C, E, and A **22. a.** sodium **b.** calcium **23. a.** B **b.** A **c.** O **d.** O **e.** A

DEFINITIONS CROSSWORD

Across
1. lipase **3.** fiber **4.** pharynx **5.** amylase **8.** anus **9.** gastric

Down:
1. liver **2.** esophagus **6.** lacteal **7.** gallbladder

CHAPTER TEST

1. b **2.** c **3.** a **4.** e **5.** f **6.** d **7.** a **8.** d **9.** e **10.** b **11.** c **12.** a **13.** b **14.** d **15.** b **16.** d **17.** b **18.** d **19.** d **20.** a **21.** b **22.** c **23.** a **24.** d **25.** d **26.** Like all other organisms, humans function at the level of the cell. Digestion occurs so that nutrients in foods can be processed to the point that they are available to individual cells. **27.** LDL refers to low-density lipoproteins—the kind that carry cholesterol away from the liver to the body where it can be deposited. This is bad for the health. HDL refers to high-density lipoproteins. These deposit cholesterol in the liver where it can be processed to bile salts and eventually eliminated from the body. **28.** The digestive system provides nutrients needed for reproduction and the growth of an unborn fetus while it develops in the uterus. Later, the newborn can be supplied with nutrients from its mother during nursing, all supplied by the digestive system.

5
COMPOSITION AND FUNCTION OF BLOOD

STUDY TIPS

Red blood cells contain the pigment hemoglobin (p. 111) that distributes oxygen throughout the body. White blood cells (p. 115) are responsible for fighting pathogens. You will learn more about the functions of white blood cells in chapter 7. Plasma (p. 118) carries nutrients as well as the proteins and platelets for producing blood clotting in response to injury and blood loss. After reading this chapter, you will more fully appreciate why you need to eat adequately to maintain a healthy internal environment. The critical functions performed for you by your blood help you survive.

When you study the mechanism of blood clotting (p. 116), make a flow diagram to help you remember what factor activates the next one (see Fig. 5.7 in the text). Draw and carefully label another diagram showing how exchange of materials occurs between the blood and tissue fluid (see Fig. 5.8 in the text). Toward the chapter's end, you will learn about the ABO system and blood typing (p. 120). This will help you to understand why it is important for medical personnel to be careful in administering transfusions.

STUDY QUESTIONS

Study the text section by section as you answer the questions that follow.

5.1 THE RED BLOOD CELLS (PAGES 111–113)

- List the characteristics of red blood cells.
- Understand how the structure of hemoglobin allows it to carry oxygen throughout the body.
- Explain how carbon monoxide prevents hemoglobin from transporting oxygen.
- Describe the life cycle of red blood cells.
- List the types of anemia and their causes.

1. Circle the items that describe hemoglobin correctly:

    a. heme contains iron
    b. globin contains iron
    c. becomes oxyhemoglobin in the tissues
    d. becomes deoxyhemoglobin in the tissues
    e. makes red blood cells red
    f. makes eosinophils red

2. The red blood cells, scientifically called a. _____, are made in the b. _____. Upon maturation, they are biconcave disks that lack a(n) c. _____ and contain d. _____. After about 120 days, red blood cells are destroyed in the e. _____ and _____. The condition of f. _____ is characterized by an insufficient number of red blood cells or not enough hemoglobin.

3. What molecule stimulates the production of red blood cells? _____

39

5.2 THE WHITE BLOOD CELLS (PAGE 115)

- List the various types of white blood cells, and describe their structures and unique functions.

4. White blood cells, scientifically called a._____, are made in the b._____.
5. Name three differences between red blood cells and white blood cells. White blood cells are a._____larger_____ in size than red blood cells; they do have a b._____nucleus_____; and they do not contain c._____.
6. Place the name of the correct white blood cell (neutrophil, eosinophil, basophil, lymphocyte, or monocyte) next to the proper description.

    a._____ An agranular cell with a large, round nucleus that occurs in two versions. The B lymphocytes produce antibodies and the T lymphocytes destroy cells that contain viruses.

    b._____ An abundant granular cell with a multilobed nucleus that phagocytizes pathogens.

    c._____ A large agranular cell that takes up residence in the tissues and differentiates into a voracious macrophage.

    d._____ A cell with blue-staining granules that takes up residence in the tissues; these become mast cells and release histamine.

    e._____ A cell with a bilobed nucleus and red-staining granules that becomes abundant during allergies and parasitic infections.

7. A type of cancer called a._____ occurs when abnormally large numbers of immature b._____ fill red bone marrow. The patient becomes both anemic and incapable of fighting disease organisms.

5.3 BLOOD CLOTTING (PAGE 116)

- Explain how clotting factors in the blood operate to produce a blood clot.

8. The following shows the reactions that create blood clots:

    platelets ───────────→ prothrombin activator
    prothrombin ─────────→ thrombin
    fibrinogen ──────────→ fibrin threads

    Does the left-hand side or the right-hand side list substances that are always present in the blood? a._____ Which substances function as enzymes? b._____ Which substance is the actual clot? c._____
9. Several nutrients are necessary for clotting to occur. Vitamin a._____ is needed for the production of prothrombin. The element b._____ is needed for the conversion of prothrombin to thrombin.

5.4 PLASMA (PAGE 118)

- Describe the composition of plasma, and give several functions for plasma proteins.

10. Plasma is mostly a._____ and b._____.
11. Place the correct plasma protein in the blank: *fibrinogen, albumin, globulin,* or *all plasma proteins.*

    a._____ transports cholesterol.

    b._____ helps blood clot.

    c._____ transports bilirubin.

    d._____ helps maintain the pH and osmotic pressure of the blood.

5.5 CAPILLARY EXCHANGE (PAGES 118–119)

- Explain how the exchange of materials occurs between blood and tissue fluid.
- Exchange of substances between blood and tissue fluid across capillary walls supplies cells with nutrients and removes wastes.

12. Label this diagram using these terms:
    - arterial end of the capillary
    - blood pressure (two times)
    - net pressure in
    - osmotic pressure (two times)
    - tissue fluid
    - venous end of the capillary
    - water and wastes
    - water, oxygen, and nutrients

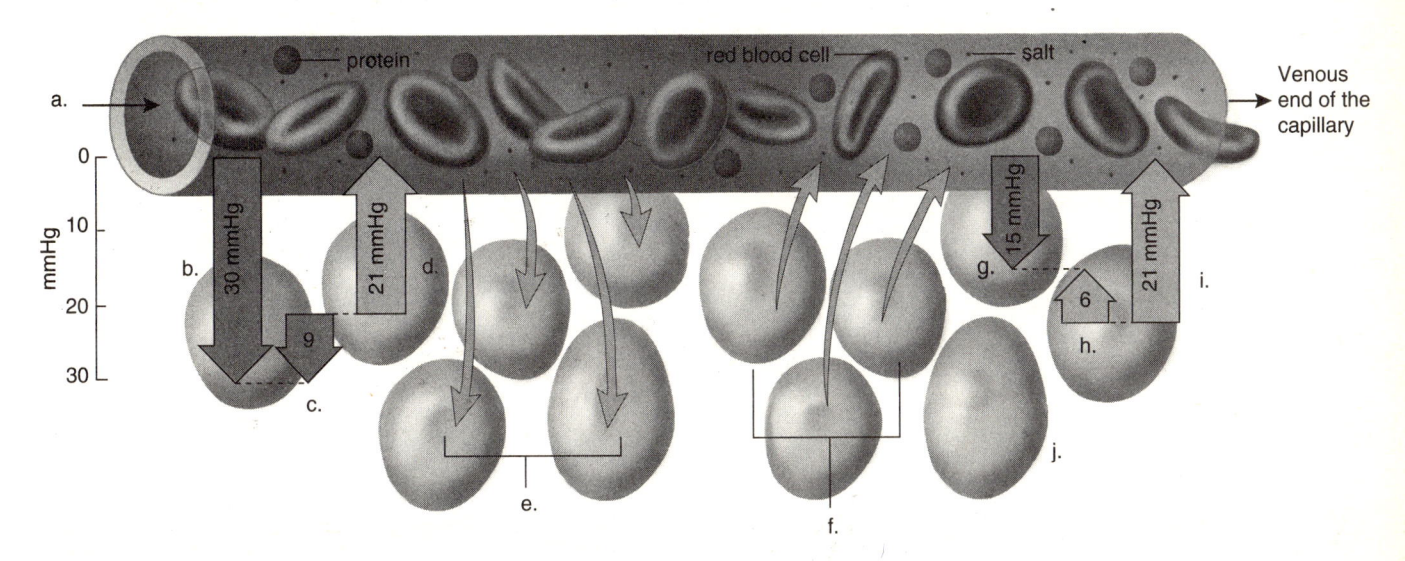

13. Explain the diagram. _____
    _____

14. Why is there excess tissue fluid, and what happens to it? _____
    _____

5.6 BLOOD TYPING (PAGES 120–121)

- Describe how antibodies and antigens determine blood type in the ABO system.
- Describe the possible Rh factor complications of pregnancy.

15. This diagram shows the results of typing someone's blood. What is the blood type? _____

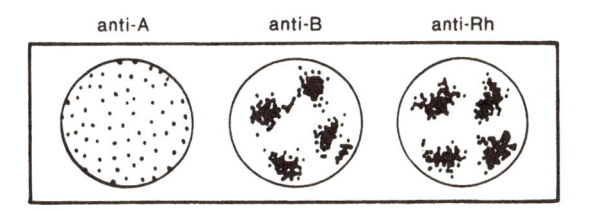

    anti-A   anti-B   anti-Rh

16. Draw a similar diagram showing the results if someone has AB negative blood.

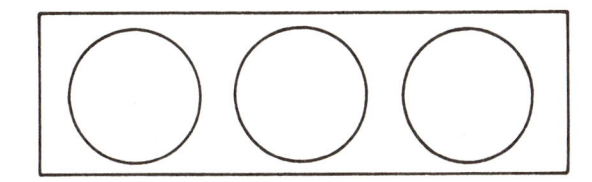

41

17. The following table indicates the blood types. Fill in the third and fourth columns by using this formula: The donor's antigen(s) must not be of the same type letter as the recipient's antibody (antibodies).

Blood Type	Antigen	Antibody	Can Receive from	Can Donate to
A	A	Anti-B	a.	b.
B	B	Anti-A	c.	d.
AB	A, B	———	e.	f.
O	None	Anti-A and B	g.	h.

18. Circle the risks associated with donating blood.
    a. Infection with HIV.
    b. Contracting hepatitis.
    c. Slight chance of feeling faint or dizzy.

19. Consider these possible combinations of mates:
    Rh⁺ mother and Rh⁻ father   Rh⁻ mother and Rh⁻ father
    Rh⁺ mother and Rh⁺ father   Rh⁻ mother and Rh⁺ father

    Which of these combinations can cause pregnancy difficulties and why?
    _____
    _____

DEFINITIONS WORDSEARCH

Review key terms by completing this wordsearch using the following alphabetized list of terms:

```
A L R P M T Y H U J F V B U M P
A N E M I A F T G D C U M O L E
R T D O P L E T G I K I D T B L
A E B O L I K A T B G U J I Y L
G I L U K I N G T F C V R M I K
C L O T T I N G Y H U R P O L I
D E O R R T I L O O P H U N I A
R E D S D R U U E R T E R Y T M
H A C X W A Z T H J N M U R E S
F T E F T U K I P L O O M L L A
A L L D F E S N Y U G G H U E L
C O L D S E T A F V C L F Y T P
T R I P D Y H T F G I O O D A Z
O R G T U N T I J I O B G Y L F
R T R L I H P O S A B I A D P Y
T H R O M B I N I G H N D O P W
```

agglutination
anemia
basophil
clotting
hemoglobin
lymph
plasma
platelet
red blood cell
Rh factor
serum
thrombin

a. _____ Fluid derived from tissue fluid that circulates in lymphatic vessels.
b. _____ Light yellow liquid left after clotting of blood.
c. _____ Condition in which shortage of hemoglobin or red blood cells leads to oxygen shortage in blood.
d. _____ Type of white blood cell that is granular and can be stained with a basic dye.

e. _____ Liquid portion of blood.
f. _____ Enzyme that triggers the conversion of fibrinogen to fibrin during blood clotting.
g. _____ Erythrocyte.
h. _____ Clumping of red blood cells during antigen-antibody reactions.
i. _____ Iron-containing protein in red blood cells that carries oxygen.
j. _____ Cell fragment in blood that functions in clotting.
k. _____ One type of antigen on red blood cells, separate from ABO system of proteins.
l. _____ Process of blood coagulation, usually when injury occurs.

CHAPTER TEST

OBJECTIVE TEST

Do not refer to the text when taking this test.

___ 1. Hemoglobin
   a. transports O_2.
   b. transports CO_2.
   c. picks up hydrogen ions.
   d. all of these

___ 2. Carbon monoxide is poisonous due to the fact that it
   a. combines more readily with hemoglobin than oxygen does.
   b. destroys red blood cells.
   c. destroys the body's tissues.
   d. destroys oxygen.

___ 3. Which of the following characterizes anemia?
   a. low red blood cell count, low hemoglobin, or both
   b. viral infection
   c. congenital disease
   d. All of these are correct.

___ 4. When blood travels through capillary beds and gives off oxygen to tissue cells, hemoglobin can now be called
   a. oxyhemoglobin.
   b. deoxyhemoglobin.
   c. carboxyhemoglobin.
   d. deaminated.

___ 5. Aged red blood cells are destroyed in the
   a. red bone marrow.
   b. lungs.
   c. lymph nodes.
   d. spleen and liver.

___ 6. Which of the following white blood cells has granules and is phagocytic?
   a. lymphocyte
   b. basophil
   c. monocyte
   d. neutrophil

___ 7. Which of these is *not* a valid contrast between red blood cells and white blood cells?
   red   white
   a. erythrocyte–leukocyte
   b. phagocytic–motile
   c. lacks nucleus–has nucleus
   d. numerous–less numerous

___ 8. Choose the best description of neutrophils.
   a. multilobed nuclei, phagocytic, do not take up stain
   b. U-shaped nucleus, dark blue after staining, turn into mast cells
   c. bilobed nucleus, red after staining, present with allergies
   d. kidney-shaped nucleus, phagocytic, turn into macrophages

___ 9. Choose the best description of basophils.
   a. lobed nuclei, phagocytic, do not take up stain
   b. U-shaped nucleus, dark blue after staining, related to mast cells
   c. bilobed nucleus, red after staining, present with allergies
   d. kidney-shaped nucleus, phagocytic, turn into macrophages

___ 10. Plasma transports
   a. nutrients.
   b. CO_2.
   c. hormones.
   d. All of these are correct.

___ 11. Which plasma protein becomes the threads of a clot?
   a. prothrombin
   b. thrombin
   c. prothrombin activator
   d. fibrinogen

___ 12. Which of these is *not* a normal function of plasma proteins?
   a. maintaining osmotic pressure
   b. a widely used source of nutrition for the body
   c. fighting infection
   d. contributing to blood clotting

___ 13. Plasma is
   a. the same as tissue fluid.
   b. the liquid remaining after blood clots.
   c. the liquid portion of the blood.
   d. All of these are correct.

___14. At a capillary
   a. oxygen is exchanged for carbon dioxide.
   b. glucose is exchanged for amino acids.
   c. water is exchanged for proteins.
   d. waste material carried in blood is deposited into tissues.
___15. At a capillary
   a. glucose and oxygen exit from the venous end, and carbon dioxide enters at the arterial end.
   b. glucose exits from the arterial end, and oxygen and carbon dioxide enter at the venous end.
   c. blood pressure increases as the cross-sectional area increases.
   d. glucose and oxygen exit from the arterial end, and carbon dioxide enters at the venous end.
___16. Water leaves capillaries at their arterial ends because
   a. osmotic pressure gradients are in opposite directions.
   b. blood pressure is greater than the osmotic pressure.
   c. a gradient is established for diffusion.
   d. osmotic pressure is always greater than blood pressure.
___17. Water reenters capillaries at their venous ends because of
   a. active transport from interstitial fluid.
   b. osmotic pressure of blood drawing fluid from tissues.
   c. increasing blood pressure.
   d. increasing hemoglobin production.
_a_18. Which mineral is needed for clotting?
   a. iron
   b. calcium
   c. thrombin
   d. manganese
_c_19. Which of these is needed for clotting?
   a. platelets
   b. vitamin K
   c. fibrinogen and prothrombin proteins
   d. All of these are correct.

_c_20. The last step in blood clotting
   a. requires potassium ions.
   b. occurs outside the bloodstream.
   c. converts thrombinogen to thrombin.
   d. converts fibrinogen to fibrin.
_c_21. Blood types are determined by the individual's
   a. antibodies.
   b. antigens.
   c. lymphocytes.
   d. phagocytes.
___22. A person with blood type O lacks
   a. antigens on the red blood cells.
   b. antigens in the plasma.
   c. the Rh factor.
   d. Both b and c are correct.
___23. The agglutination of red blood cells occurs whenever
   a. appropriate antibodies bind with antigens on red blood cells.
   b. a person receives a blood transfusion from someone with an incompatible blood type.
   c. complementary antibodies combine.
   d. blood cells are destroyed by leukocytes.
   e. Both a and b are correct.
___24. In which case will hemolytic disease of the newborn occur?
   a. Rh⁻ mother; Rh⁺ fetus
   b. Rh⁺ mother; Rh⁻ fetus
   c. A mother; O fetus
   d. Both a and b are correct.
___25. An Rh⁺ fetus being carried by an Rh⁻ mother
   a. develops antibodies to the mother's blood.
   b. develops antigens to the mother's blood.
   c. may have its red blood cells attacked by antibodies made by the mother.
   d. may have its red blood cells attacked by antigens made by the mother.

THOUGHT QUESTIONS

Answer in complete sentences.
26. What concerns should a person have when donating blood?

27. What do you need to know about a person's plasma and red blood cells before deciding who can donate blood to whom?

Test Results: _____ number correct ÷ 27 = _____ × 100 = _____%

Answer Key

Study Questions

1. a, d, e **2. a.** erythrocytes **b.** red bone marrow **c.** nucleus **d.** hemoglobin **e.** liver, spleen **f.** anemia **3.** erythropoietin **4. a.** leukocytes **b.** red bone marrow **5. a.** larger **b.** nucleus **c.** hemoglobin **6. a.** lymphocyte **b.** neutrophil **c.** monocyte **d.** basophil **e.** eosinophil **7. a.** leukemia **b.** white blood cells (leukocytes) **8. a.** left-hand side **b.** prothrombin activator and thrombin **c.** fibrin threads **9. a.** K **b.** calcium **10. a.** water **b.** plasma proteins **11. a.** globulin **b.** fibrinogen **c.** albumin **d.** all plasma proteins **12. a.** arterial end of capillary **b.** blood pressure **c.** net pressure out **d.** osmotic pressure **e.** water, oxygen, and nutrients **f.** water and wastes **g.** blood pressure **h.** net pressure in **i.** osmotic pressure **j.** tissue fluid. **13.** At the arterial end of a capillary, blood pressure is higher than osmotic pressure; therefore water, nutrients, and oxygen leave a capillary. At the venous end of a capillary, osmotic pressure is higher than blood pressure; therefore water (and wastes) enters a capillary. In this way tissue fluid is refreshed. **14.** This system never retrieves all the water that leaves capillaries, and excess tissue fluid is picked up by lymphatic vessels and returned to the bloodstream. **15.** B$^+$ **16.** clumping should occur for anti-A and anti-B. **17. a.** A, O **b.** A, AB **c.** B, O **d.** B, AB **e.** A, B, AB, O **f.** AB **g.** O **h.** A, B, AB, O **18.** c **19.** Rh$^-$ mother and Rh$^+$ father, because the mother might form antibodies to destroy red blood cells of this or a future baby who is Rh$^+$.

Definitions Wordsearch

```
              R
A N E M I A
              D                              L
              B         A                    Y
              L         G           M
C L O T T I N G                 P
              O         L           H        A
R             D         U           E    T   M
H             C         T           M U R E S
F             E         I           O        L A
A             L         N           G    E   L
C             L         A           L    T   P
T                       T           O    A
O                       I           B    L
R         L I H P O S A B I              P
T H R O M B I N              N
```

a. lymph **b.** serum **c.** anemia **d.** basophil **e.** plasma **f.** thrombin **g.** red blood cell **h.** agglutination **i.** hemoglobin **j.** platelet **k.** Rh factor **l.** clotting

Chapter Test

1. d **2.** a **3.** a **4.** b **5.** d **6.** d **7.** b **8.** a **9.** b **10.** d **11.** d **12.** b **13.** c **14.** a **15.** d **16.** b **17.** b **18.** b **19.** d **20.** d **21.** b **22.** a **23.** e **24.** a **25.** c **26.** There is little to be concerned about regarding your safety as a donor; however, you should always be aware of your own health status and whether you carry diseases that would harm the recipient of your blood. Some people feel faint or dizzy for a short while after donating blood. **27.** To decide who can give blood to whom, you need to know the antigens on the red blood cells of the potential donors and the antibodies in the plasma of the potential recipients.

6

CARDIOVASCULAR SYSTEM

STUDY TIPS

Learning about the anatomy and physiology of the heart is one of the first important steps one can take toward keeping healthy and fit. In this chapter, you will learn about the cardiovascular system, including the blood vessels, heart, pathways of circulation, and circulatory disorders.

When you study the pathway of circulation through the heart (p. 129), remember that blood returning to the heart must next be sent to the lungs to pick up oxygen. The heart is slightly off center to the left side of the chest. That means the right side of the heart is closer to the midline *and* closer to the middle of the lungs. Thus, it is the right side of the heart that pumps blood to the lungs. Oxygenated blood returns from the lungs to the top left of the heart. The bottom left side of the heart, with its thicker muscular walls, pumps oxygenated blood to the remainder of the body. Also, remember that blood flows first into the atria (A), then into the ventricles (V) within the heart (A comes before V in the alphabet).

STUDY QUESTIONS

Study the text section by section as you answer the questions that follow.

6.1 BLOOD VESSELS (PAGES 126–127)

- Name and describe the structure and function of arteries, capillaries, and veins.
- Explain how blood flow may bypass certain capillary beds.

1. Label the blood vessels in this diagram, using the alphabetized list of terms.
    arterioles
    artery
    capillaries
    heart
    vein
    venules

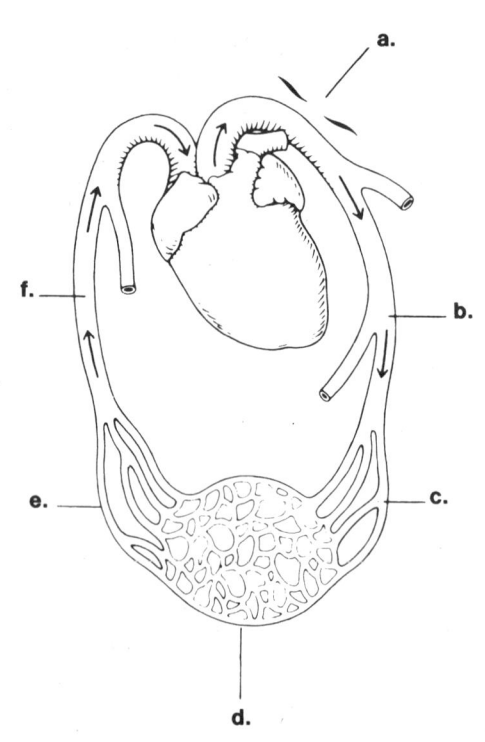

46

2. Match the vessels in the key to the appropriate statements.
   Key artery vein capillary

   a. ~~capillary~~ artery _____ Has the thickest walls.
   b. vein _____ Has valves.
   c. artery _____ Takes blood away from the heart.
   d. vein _____ Takes blood to the heart.
   e. capillary _____ Exchanges carbon dioxide and oxygen with tissues.
   f. vein _____ Nervous stimulation causes these to constrict during hemorrhaging; these also act as a blood reservoir.

3. Explain how it is possible to bypass capillary beds by shunting blood directly from arteriole to venule.
   _____
   _____

6.2 THE HEART (PAGES 128–131)

- Name the parts of the heart and their functions.
- Trace the path of blood flow through the heart.
- Explain how the conduction system of the heart controls the heartbeat.
- Label and explain a normal electrocardiogram.
- Describe how impulses from the nervous system and hormones can modify the heart rate.

4. Trace the path of blood through the heart from the vena cava to the lungs. a. Right Atrium → Right Ventricle → A.V. → Pulmonary Arteries → P.C.B → Left Atery

   from the lungs to the aorta. b. Left Atery → Left Ventricle → AV → Aorta →
   _____

5. Label the parts of the heart using the alphabetized list of terms.

   aorta
   aortic semilunar valve
   atrioventricular (mitral) valve
   atrioventricular (tricuspid) valve
   AV node
   chordae tendineae
   inferior vena cava
   left atrium
   left ventricle
   pulmonary artery
   right atrium
   right ventricle
   SA node
   septum
   superior vena cava

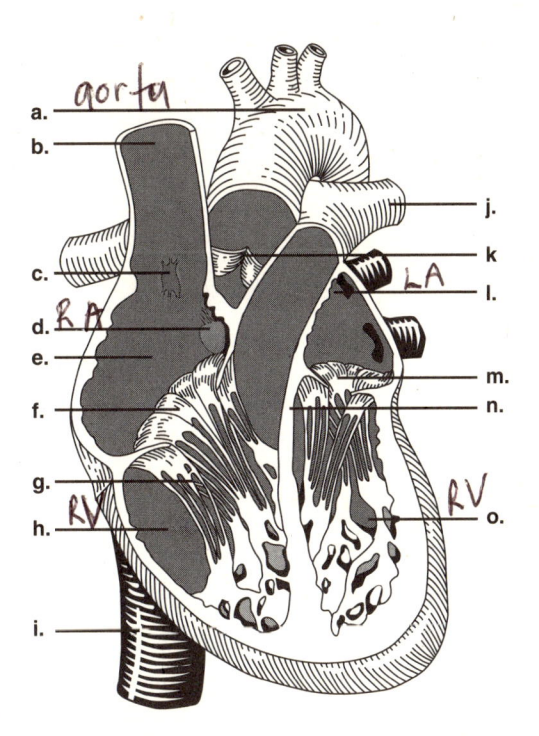

   a. aorta
   b. ___
   c. ___
   d. RA
   e. ___
   f. ___
   g. ___
   h. RV
   i. ___
   j. ___
   k. ___
   l. LA
   m. ___
   n. ___
   o. RV

6. How does the thickness of the walls of the ventricles relate to their functions? _____
   _____

7. Referring to Figure 6.6 in the text, fill in the following table with the words *systole* (contraction) and *diastole* (relaxation) to show what happens during the 0.85 seconds of one heartbeat.

time	Atria	Ventricles
0.15 sec	a.	d.
0.30 sec	b.	e.
0.40 sec	c.	f.

8. Heart sounds. When the atria contract, this forces blood through the a. __AV__ valves into the chambers called the b. __Right ventricle__. The closing of these valves is the lub sound. Next, the ventricles contract and force the blood into the arteries. Now the c. __semilunar__ valves close, making the dub sound.

9. Match the following phrases to this key.
   Key SA node   AV node
   a. __SA__ pacemaker
   b. _____ contraction of ventricles
   c. _____ base of right atrium near the septum
   d. _____ Purkinje fibers

10. Electrocardiogram. Of what significance is each of the following?
    P wave a._____
    QRS wave b._____
    T wave c._____

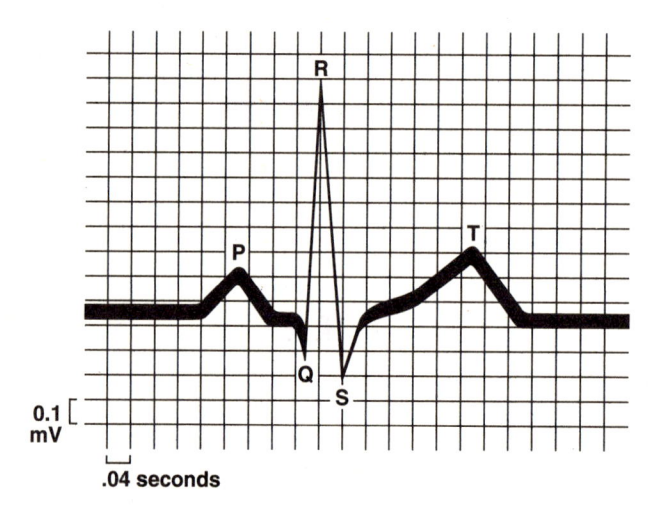

11. Match the following actions to the divisions of the nervous system.
    Key parasympathetic system   sympathetic system
    a. __parasympathetic__ normal body functions
    b. __sympathetic__ active under times of stress
    c. __sympathetic__ releases norepinephrine to speed up heart
    d. __parasympathetic__ slows heart rate

12. Does the hormone thyroxine from the thyroid speed or slow the heart rate? a._____ Does the adrenal gland hormone, epinephrine, speed or slow the heart rate? b._____

48

6.3 FEATURES OF THE CARDIOVASCULAR SYSTEM (PAGES 132–133)

- Identify the factors that influence blood pressure.

Questions 13–17 are based on this diagram.

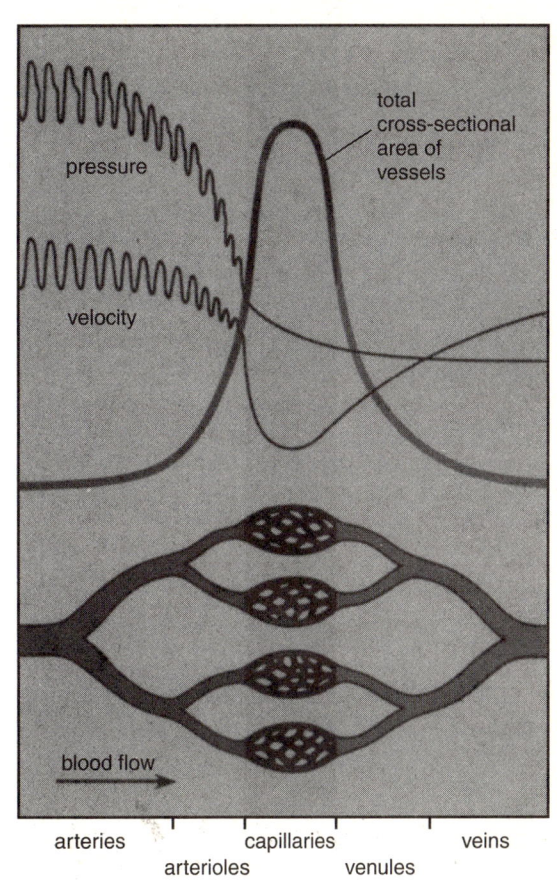

13. What force accounts for blood flow in arteries? __antrial systol__

14. Why does this force fluctuate? _____

15. What causes the pressure and velocity to drop off? _____

16. What force accounts for blood flow in the veins? __diastol systol__

17. What keeps blood from flowing backwards in veins? __ventricle systol__

6.4 THE VASCULAR PATHWAYS (PAGES 134–135)

- Describe the pulmonary circuit of circulation.
- Identify the major vessels of the systemic circuit.

18. Trace the path of blood to the lungs:
   right ventricle
   a. _____
   lungs
   b. _____
   left atrium

   from the legs:
   legs
   c. _____
   d. _____
   right atrium

49

19. Trace the path of blood from the aorta to the liver:　　from the liver:
    aorta　　　　　　　　　　　　　　　　　　　　　　　　　liver

    a._____　　　　　　c._____
    digestive tract　　　　　　　　　　　　　　　　　　　vena cava

    b._____
    liver

6.5 Cardiovascular Disorders (pages 136–139)
- Discuss the factors that can lead to cardiovascular disease and how they can be prevented.

20. Match the items in the key to the statements that follow.
    Key artificial pacemaker needed　dietary restriction of salt and/or cholesterol　donor heart transplant
    coronary bypass

    Clearing clogged arteries was unsuccessful.　a._____

    Blood pressure is 200/140.　b._____

    Heartbeat is irregular.　c._____

    Congestive heart failure is present.　d._____

21. Match the items in the key to the phrases that follow.
    Key thrombus and embolus　atherosclerosis and hypertension　varicose veins　hemorrhoids

    stroke and heart attack　a._____

    varicose veins in rectum　b._____

    weakened valves　c._____

6.6 Homeostasis (page 140)
- Discuss how the circulatory system works with other systems of the body to maintain homeostasis.

22. The *Working Together* diagram in your textbook indicates ways in which the circulatory system benefits other organ systems of the body. Match the organ systems with the correct descriptions.

    a. Blood vessels transport leukocytes and antibodies; blood services lymphoid organs and is source of tissue fluid that becomes lymph.
    b. Blood vessels deliver nutrients and oxygen to bones; carry away wastes.
    c. Blood vessels deliver nutrients and oxygen to muscles; carry away wastes.
    d. Blood vessels deliver wastes to be excreted; blood pressure aids kidney function; blood services urinary organs.
    e. Blood vessels deliver nutrients and oxygen to skin, carry away wastes; blood clots if skin is broken.
    f. Blood vessels transport hormones from glands; blood services glands; heart produces atrial natriuretic hormone.
    g. Blood vessels transport nutrients from digestive tract to body; blood services digestive organs.
    h. Blood vessels transport sex hormones; vasodilation causes genitals to become erect; blood services reproductive organs.
    i. Blood vessels deliver nutrients and oxygen to neurons; carry away wastes.
    j. Blood vessels transport gases to and from lungs; blood services respiratory organs.

    _____ integumentary system
    _____ respiratory system
    _____ lymphatic system
    _____ muscular system
    _____ skeletal system
    _____ nervous system
    _____ urinary system
    _____ reproductive system
    _____ endocrine system
    _____ digestive system

Definitions Wordsearch

Review key terms by completing this wordsearch using the following alphabetized list of terms:

```
M Y O C A R D I U M L I M
O U I L T R G H D C V U N
Q R G T R G J K I N D E I
I C A P I L L A R Y R E E
R A T R O A L O P E R V V
D E C V V S G E T H E L Y
E Y S D E I Y X Z E R Y R
E R N J N H U S J L O P A
L E G U T A S D T A N D N
O T G B R M H K I O G O O
T R D E I N U N S U L L M
S A G Y C G A Y D O T E L
A Y S E U S A I D F U G U
I R R T L D V S K O P L P
D A D A A R T E R I O L E
M N A M R A W P I L L L A
U O R A V J I T M J E N N
I R F E A R M U A E D E A
N O E I L C H M A R D M A
E C M I V M I C O T T O N
A L C H E C H R W E N D Y
```

aorta — h
arteriole — b
atrioventricular valve — a
capillary — i
coronary artery — j
diastole — f
myocardium — c
pulmonary vein — g
septum — d
systole — e

a. _arteriole_ ✗ Valve located between the atrium and the ventricle.
b. _aorta_ ✗ Vessel that takes blood from an artery to capillaries.
c. _myocardium_ ✓ Cardiac muscle in the wall of the heart.
d. _septum_ Partition in heart that divides it into left and right halves.
e. _systole_ Contraction of a heart chamber.
f. _diastole_ Relaxation of a heart chamber.
g. _pulmonary vein_ Blood vessel returning from lungs to heart.
h. _arteriole_ ✗ Large artery leaving heart with blood from the left ventricle.
i. _capillary_ Microscopic vessel connecting arterioles to venules.
j. _coronary artery_ Artery that supplies blood to the wall of the heart.

Chapter Test

Objective Test

Do not refer to the text when taking this test.

c 1. Which type of blood vessel allows exchange of material between the blood and the tissues?
   a. arteries
   b. arterioles
   c. capillaries ✓
   d. veins
   e. venules

b 2. What is the function of the heart valves?
   a. to push blood
   b. to prevent backflow ✓
   c. to stimulate the heart
   d. to give support to the heart

51

_a_ 3. Arteries
   a. carry blood away from the heart.
   b. carry blood toward the heart.
   c. have valves.
   d. Both a and b are correct.

_c_ 4. Shunting of blood is possible because of a thoroughfare channel that joins _____ to _____ directly.
   a. arterioles; venules
   b. venules; capillaries
   c. arteries; veins
   d. arterioles; veins

_b_ 5. Which of these vessels have thinner walls?
   a. arteries
   b. veins
   c. Both are the same.

_a_ 6. The venae cavae
   a. carry blood to the right atrium.
   b. carry blood away from the right atrium.
   c. join with the aorta.
   d. have a high blood pressure.

_f_ 7. The chamber of the heart that receives blood from the pulmonary veins
   a. is the right atrium.
   b. is the left atrium.
   c. contains oxygenated blood.
   d. contains deoxygenated blood.
   e. a and c
   f. b and c
   g. b and d

_a_ 8. The coronary arteries carry blood
   a. from the aorta to the heart tissues.
   b. from the heart to the brain.
   c. directly to the heart from the pulmonary circuit.
   d. from the lungs directly to the left atrium.

_d_ 9. Which of these chambers has the thickest walls?
   a. right atrium
   b. right ventricle
   c. left atrium
   d. left ventricle

_f_ 10. When the atria are contracting, the ventricles are
   a. contracting.
   b. relaxing.
   c. in diastole.
   d. in systole.
   e. a and c
   f. b and c
   g. b and d

_a_ 11. The SA node
   a. works only when it receives a nerve impulse.
   b. is located in the left atrium.
   c. initiates the heartbeat.
   d. All of these are correct.

_a_ 12. The first wave (the _P_ wave) of an ECG represents
   a. atrial contraction.
   b. ventricular contraction.
   c. ventricular relaxation.
   d. atrial relaxation.

_b_ 13. The heart sounds are due to
   a. blood flowing.
   b. the closing of the valves.
   c. the heart muscle contracting.
   d. blood pressure in the aorta.

_c_ 14. Blood flows in veins because of
   a. contraction of valves.
   b. arterial blood pressure.
   c. capillary blood pressure.
   d. skeletal muscle contraction.

_c_ 15. Systole refers to the contraction of the
   a. major arteries.
   b. SA node.
   c. atria and ventricles.
   d. major veins.

_b_ 16. Blood pressure falls off drastically in the capillaries because the capillaries
   a. contain valves.
   b. become veins.
   c. have a large cross-sectional area.
   d. All of these are correct.

_c_ 17. Which of these correctly traces the path of blood from the left ventricle to the head?
   a. left ventricle, subclavian artery, head
   b. left ventricle, pulmonary artery, head
   c. left ventricle, aorta, carotid artery, head
   d. left ventricle, vena cava, jugular vein, head

_b_ 18. Blood pressure
   a. is the same in all blood vessels.
   b. is highest in the aorta.
   c. is measured by taking an ECG.
   d. never rises above normal.

_a_ 19. The major portion of the circulatory system is called the
   a. systemic circuit.
   b. pulmonary circuit.
   c. hepatic portal circuit.
   d. coronary circuit.

_c_ 20. Blood flowing to the lungs leaves the heart via the _____ and returns to the heart via the _____.
   a. aorta; superior vena cava
   b. superior vena cava; aorta
   c. pulmonary arteries; pulmonary veins
   d. aorta; pulmonary veins

____ 21. The jugular vein carries blood from the
   a. head.
   b. vena cava.
   c. arm.
   d. aorta.

52

___22. Blood moves slowly in capillaries
  a. which facilitates tissue exchange.
  b. because they have valves.
  c. because they have thick walls.
  d. because venules are smaller than capillaries.

___23. People with atherosclerosis often experience
  a. high blood pressure.
  b. a heart attack.
  c. a thrombus.
  d. a stroke.
  e. Any of these is correct.

___24. A heart attack is due to a blocked
  a. pulmonary artery.
  b. coronary artery.
  c. aorta.
  d. vena cava.

___25. Phlebitis and hemorrhoids are conditions involving
  a. arteries.
  b. capillaries.
  c. veins.
  d. arterioles.

THOUGHT QUESTIONS

Answer in complete sentences.

26. Why can arteries expand without rupturing?

27. Explain how the digestive system and circulatory system benefit each other.

Test Results: _____ number correct ÷ 27 = _____ × 100 = _____%

53

Answer Key

Study Questions

1. a. heart **b.** artery **c.** arterioles **d.** capillaries **e.** venules **f.** vein   **2. a.** artery **b.** vein **c.** artery **d.** vein **e.** capillary **f.** vein   **3.** The shunting of blood around capillary beds is possible because each bed has a thoroughfare channel that allows blood to flow directly from arteriole to venule. Sphincter muscles prevent blood from flowing into the capillaries.   **4. a.** vena cava, right atrium, atrioventricular valve, right ventricle, pulmonary semilunar valve, pulmonary artery, lungs **b.** lungs, pulmonary veins, left atrium, atrioventricular valve, left ventricle, aortic semilunar valve, aorta   **5. a.** aorta **b.** superior vena cava **c.** SA node **d.** AV node **e.** right atrium **f.** atrioventricular (tricuspid) valve **g.** chordae tendineae **h.** right ventricle **i.** inferior vena cava **j.** pulmonary artery **k.** aortic semilunar valve **l.** left atrium **m.** atrioventricular (mitral) valve **n.** septum **o.** left ventricle   **6.** The left ventricle is thicker-walled than the right ventricle because the left one must pump blood the greater distance to the entire body. The right ventricle only pumps the shorter distance to the lungs.   **7. a.** systole **b.** diastole **c.** diastole **d.** diastole **e.** systole **f.** diastole   **8. a.** atrioventricular **b.** ventricles **c.** semilunar   **9. a.** SA node **b.** AV node **c.** AV node **d.** AV node   **10. a.** associated with atrial systole **b.** associated with ventricular systole **c.** associated with ventricular recovery   **11. a.** parasympathetic system **b.** sympathetic system **c.** sympathetic system **d.** parasympathetic system   **12. a.** speeds **b.** speeds   **13.** blood pressure   **14.** systole and diastole of the left ventricle of the heart   **15.** distance from heart and increase in cross-sectional area of blood vessels   **16.** skeletal muscle contraction   **17.** valves   **18. a.** pulmonary arteries **b.** pulmonary veins **c.** iliac vein **d.** inferior vena cava   **19. a.** mesenteric arteries **b.** hepatic portal vein **c.** hepatic vein   **20. a.** coronary bypass **b.** dietary restriction **c.** artificial pacemaker needed **d.** donor heart transplant   **21. a.** thrombus and embolus; atherosclerosis and hypertension **b.** hemorrhoids **c.** varicose veins   **22.** e, j, a, c, b, i, d, h, f, g

Definitions Wordsearch

```
M Y O C A R D I U M
        T               N
        R               I
    C A P I L L A R Y   E
    A T R O A           V
        V S             Y
    Y   E Y             R
  E R   N     S         A
  L E   T         T     N
  O T   R             O M
  T R   I           L   
  S A   C           E L U
  A Y   U               P
  I R   L     S         
  D A   ARTERIOLE       
  N     R   P           
  O     V   T           
  R     A   U           
  O     L   M           
  C     V               
        E               
```

a. atrioventricular valve **b.** arteriole **c.** myocardium **d.** septum **e.** systole **f.** diastole **g.** pulmonary vein **h.** aorta **i.** capillary **j.** coronary artery

Chapter Test

1. c **2.** b **3.** a **4.** a **5.** b **6.** a **7.** f **8.** a **9.** d **10.** f **11.** c **12.** a **13.** b **14.** d **15.** c **16.** c **17.** c **18.** b **19.** a **20.** c **21.** a **22.** a **23.** e **24.** b **25.** c **26.** The wall of an artery has a middle layer of elastic tissue and smooth muscle and an outer layer of fibrous connective tissue. This arrangement allows for both strength and flexibility.   **27.** The circulatory system transports nutrients from the digestive tract to the rest of the body and also services the organs of the digestive tract. The digestive tract provides nutrients for blood cell formation and the formation of plasma proteins. The liver detoxifies the blood, makes plasma proteins, and destroys old red blood cells.

7

LYMPHATIC SYSTEM AND IMMUNITY

STUDY TIPS

How the body defends itself from disease is one of the most fascinating areas of human biology. In this chapter, you will learn about the organs, cells, and special molecules involved in the lymphatic system and immunity.

Compare the structure of lymphatic vessels (p. 146) with those of the circulatory system from chapter 6. In your notes, make a list of each immune system cell type (pp. 150–155) and its specific set of functions. Do the same for the molecules of the immune system: interferon (p. 150), complement (p. 150), lymphokines, and so forth. By doing some careful "bookkeeping" of terms in this chapter, you will not confuse them with each other. Understand as well that the HLA are self antigens (pp. 154–155), a way for cells of the immune system to recognize each other along with the other cells of your body. Only when the immune system cells have an accurate inventory of the "allies" in your body can it readily recognize and attack the "enemies."

STUDY QUESTIONS

Study the text section by section as you answer the questions that follow.

7.1 LYMPHATIC SYSTEM (PAGES 146–148)

- Describe the structure of lymphatic vessels.
- List three functions of the lymphatic system.
- Explain the structure and purpose of lymph nodes.
- Discuss the functions of the spleen and of the white and red pulp it contains.
- Describe how the thymus gland and red bone marrow participate in immunity.

1. Give three functions of the lymphatic system.
   a. _____
   b. _____
   c. _____

2. Indicate whether the following statements about the structure/function of lymphatic vessels and lymphoid organs are true (T) or false (F):
   _____ a. Bone marrow lacks lymphoid tissue.
   _____ b. Lymph lobules are subdivided into sinus-containing nodes.
   _____ c. The contraction of skeletal muscles blocks the return of lymph to the bloodstream.
   _____ d. The sinuses of the spleen are filled with lymph.
   _____ e. Lymphatic vessels are similar to systemic veins.
   _____ f. Lymphatic vessels contain valves.

3. Indicate whether the statements that follow are true (T) or false (F). Rewrite any false statements to make them true.
   _____ a. Vessels of the lymphatic system begin with systemic capillaries. Rewrite: _____
   _____ b. Lymph most closely resembles arterial blood. Rewrite: _____
   _____ c. The right thoracic duct serves the lower extremities, abdomen, one arm, and one side of the head and neck. Rewrite: _____
   _____ d. Lymphatic capillaries merge directly to form a particular lymphatic duct. Rewrite: _____

4. Label each of the following statements as describing the thymus (T), spleen (S), or red bone marrow (RBM):
   _____ a. contains red pulp and white pulp
   _____ b. contains stem cells
   _____ c. is located along the trachea
   _____ d. is located in the upper left abdominal cavity
   _____ e. produces hormones believed to stimulate the immune system
   _____ f. contains sinuses filled with blood
   _____ g. site of origin for all types of blood cells

7.2 NONSPECIFIC DEFENSES (PAGES 148–150)

- Describe barriers to entry, the inflammatory reaction, natural killer cells, and protective proteins as agents of nonspecific immunity.

5. Match each of the descriptions with the following defense mechanisms:
   1 barrier to entry
   2 inflammatory response
   3 complement protein
   4 natural killer cells
   _____ a. accompanied by swelling and redness
   _____ b. cilia action in the respiratory tract
   _____ c. produces holes in bacterial cell walls
   _____ d. stomach secretions
   _____ e. histamine increases capillary permeability
   _____ f. injured cells release bradykinin
   _____ g. inhabitation of normal body bacteria
   _____ h. interferon action
   _____ i. neutrophils and macrophages carry out phagocytosis
   _____ j. secretions of the oil, or sebaceous, glands
   _____ k. kills cells infected with a virus and tumor cells

7.3 SPECIFIC DEFENSES (PAGES 150–155)

- Compare the origin, maturation, and function of B cells and T cells.
- Describe the general structure of an antibody and state a function for the variable and constant regions.
- Explain how and where B cells undergo clonal selection and expansion.
- Tell how a T cell recognizes an antigen.
- List the different types of T cells and give the action of each.

6. Fill in this table with *yes* or *no:*

	Cytotoxic T Cell	B Cell
Ultimately derived from stem cells in bone marrow	a.	b.
Pass through thymus	c.	d.
Carry receptors on membrane	e.	f.
Cell-mediated immunity	g.	h.
Antibody-mediated immunity	i.	j.

B Cells and Antibody-Mediated Immunity (Page 151)

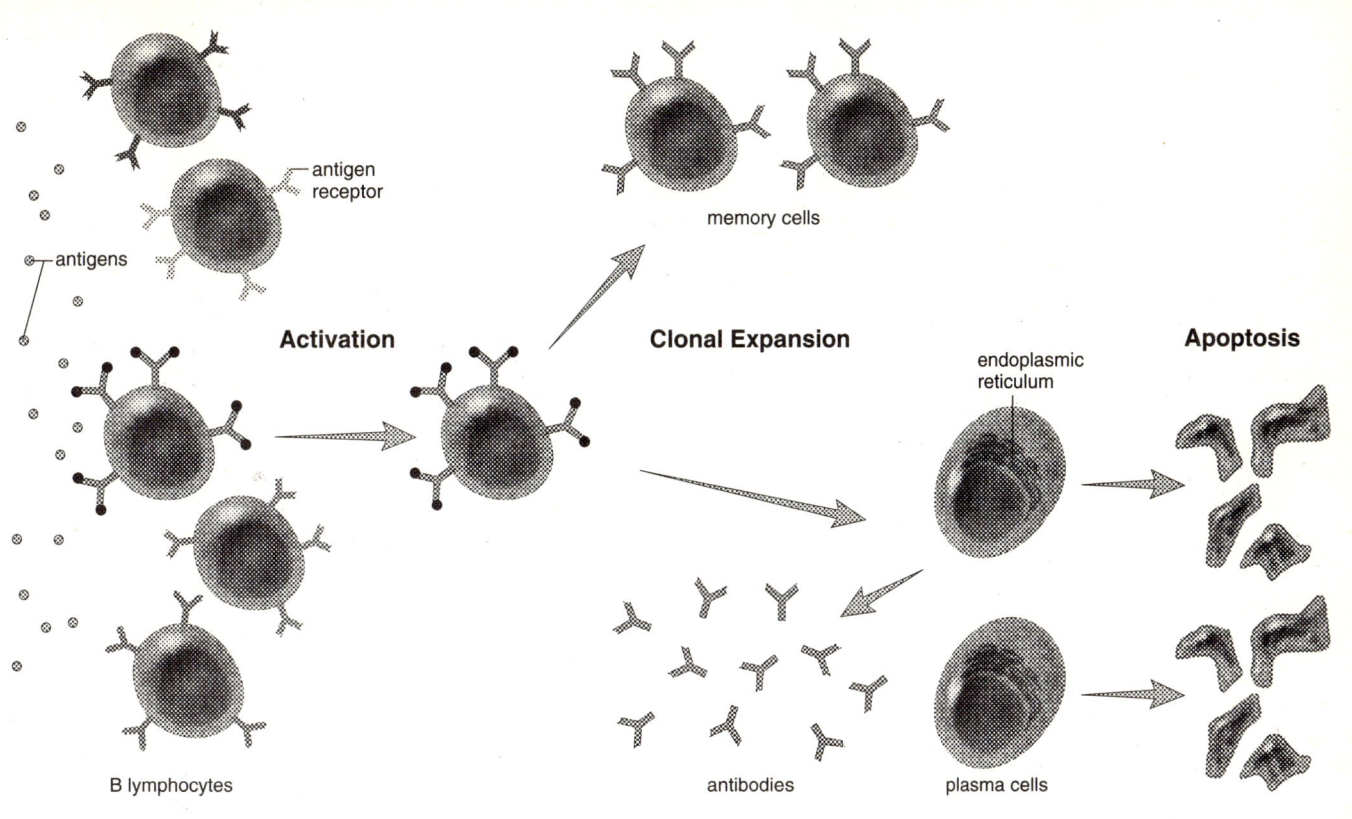

Study the diagram and answer questions 7–10.

7. Which of these makes a particular B cell undergo clonal expansion?
   a. when the correct antigen is present
   b. when any antigen is present

8. Explain the expression clonal selection theory. _____
   _____

9. Which of these are cells that result from the clonal expansion of B cells?
   a. plasma cells
   b. memory cells
   c. both types of cells.

10. What happens to these cells once the infection is under control?
    a. memory cells _____
    b. plasma cells _____

11. Label this diagram of an antibody molecule with the following terms: *antigen-binding site, constant region, heavy chain, light chain,* and *variable region.*

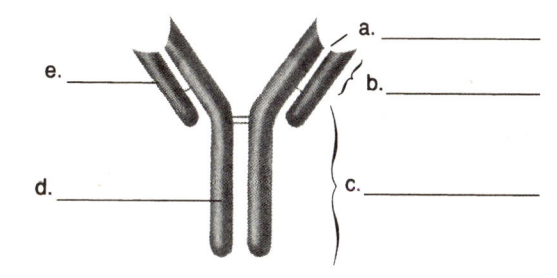

    a. _____
    b. _____
    c. _____
    d. _____
    e. _____

    What is the function of antibodies? f. _____

T Cells and Cell-Mediated Immunity (Pages 154–155)

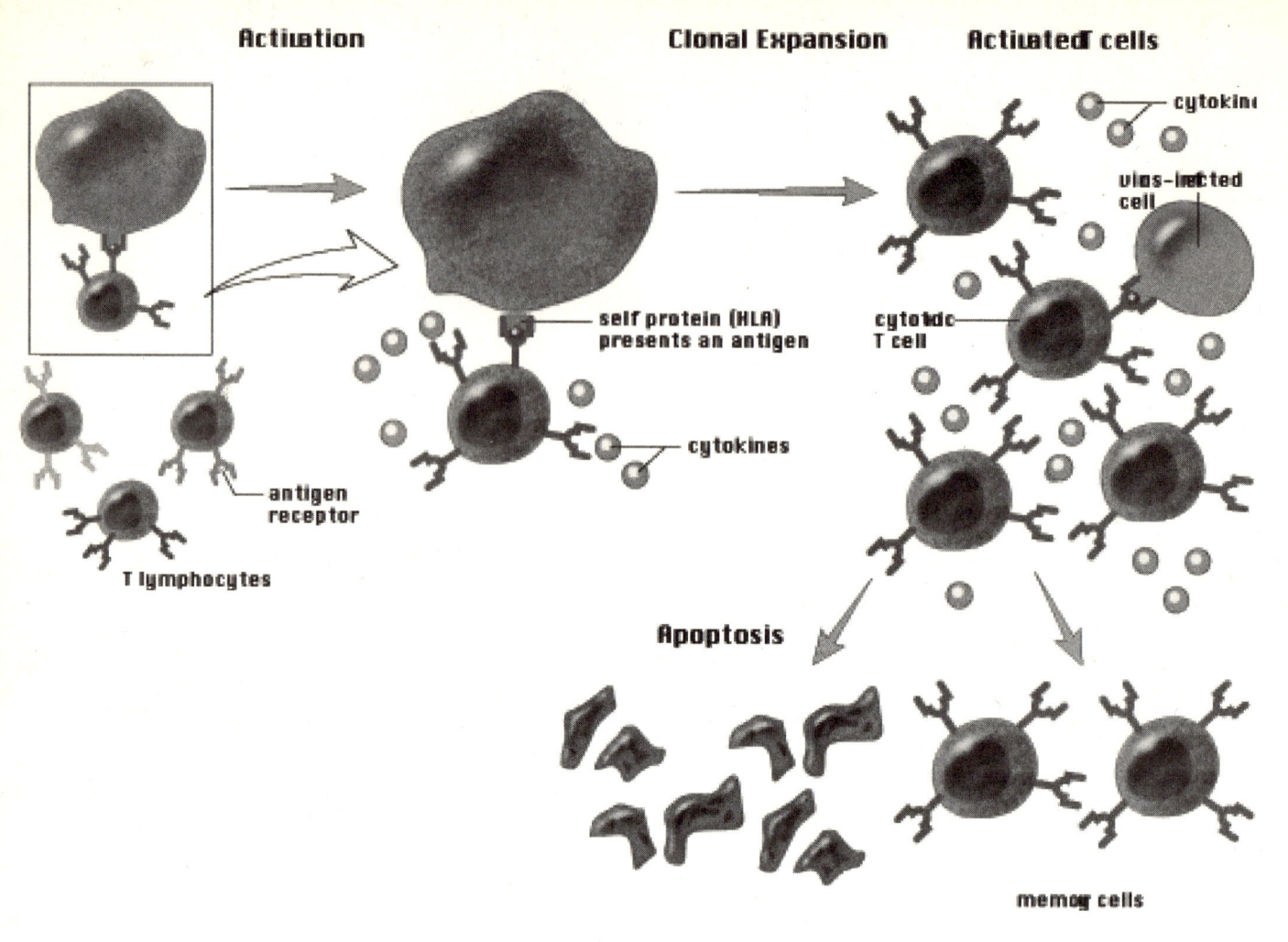

Activation — T lymphocytes, antigen receptor

Clonal Expansion — self protein (HLA) presents an antigen, cytokines

Activated cells — cytokine, virus-infected cell, cytotoxic T cell

Apoptosis

memory cells

Study the diagram and answer questions 12–14.

12. Which of these makes a particular T cell undergo clonal expansion?
    a. when the antigen is presented to the T cells by an APC
    b. when a T cell encounters an antigen

13. What is the significance of HLA antigens? a. _____
    _____

    Why are they called antigens? b. _____
    _____

14. What happens after a helper T cell recognizes an antigen? a. _____
    _____

    What happens after a cytotoxic T cell recognizes an antigen? b. _____
    _____

    What happens to T cells (except for memory T cells) after the infection is past? c. _____
    _____

7.4 Induced Immunity (pages 156–158)

- Differentiate between active and passive immunity.

15. a. Label the following diagram, using the alphabetized list of terms: *antibody titer in serum, exposure to vaccine* (needed twice), *primary response, secondary response, time (days)*

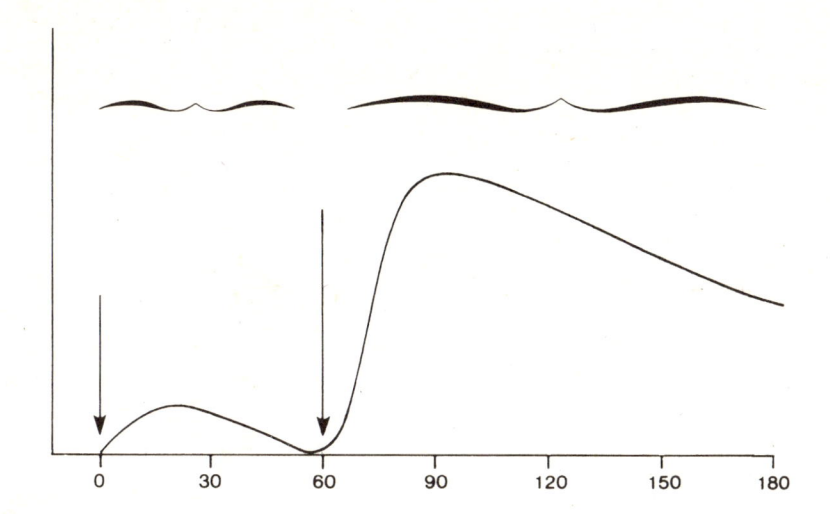

A good secondary response can be related to the b. _____ dependent on the number of
c. _____ and d. _____ cells capable of responding to a particular antigen.

16. When an individual receives antibodies from another, as when a baby breast-feeds, it is called _____.

17. List five diseases for which children should be immunized. _____
_____

18. Why is it better to immunize children in advance than to wait until the disease has been contracted?
_____

19. Name two types of cytokines, and tell what role they play in immunotherapy.
   a. _____
   b. _____

20. What are monoclonal antibodies? _____
_____

7.5 Immunity Side Effects (pages 158–160)

- List three types of immunological side effects and relate them to the function of the immune system.

21. Relate the immune response to each of these.
   allergy a. _____
   tissue rejection b. _____
   autoimmune disease c. _____

7.6 HOMEOSTASIS (PAGE 160)

- Discuss how the lymphatic and immune systems work with other systems of the body to maintain homeostasis.

22. The *Working Together* diagram in your textbook indicates ways in which the lymphatic system benefits from the actions of the other organ systems of the body. Match the organ systems with the correct descriptions.

    a. Tonsils and adenoids occur along respiratory tract; breathing aids lymph flow; lungs carry out gas exchange.
    b. Microglial cells engulf and destroy pathogens.
    c. Sex hormones influence immune functioning; acidity of vagina helps prevent pathogen invasion of body; milk passes antibodies to newborn.
    d. Skin serves as a barrier to pathogen invasion; protects lymphatic vessels.
    e. Red bone marrow produces leukocytes involved in immunity.
    f. Digestive tract provides nutrients for lymphoid organs, stomach acidity prevents pathogen invasion of body.
    g. Thymus is necessary for maturity of T lymphocytes.
    h. Blood vessels transport leukocytes and antibodies; blood services lymphoid organs and is source of tissue fluid that becomes lymph.
    i. Kidneys control volume of body fluids, including lymph.
    j. Skeletal muscle contraction moves lymph; physical exercise enhances immunity.

    _____ integumentary system
    _____ respiratory system
    _____ circulatory system
    _____ muscular system
    _____ skeletal system
    _____ nervous system
    _____ urinary system
    _____ reproductive system
    _____ endocrine system
    _____ digestive system

DEFINITIONS WORDSEARCH

Review key terms by completing this wordsearch, using the following alphabetized list of terms:

```
L E N I M A T S I H E R T M
I G S D C O M P L E M E N T
E A N T I B O D Y D L P A O
N H U I P R K J N E G H U D
I P H L L A R N O M J U T V
K O C E A D H E R A L O O B
O R M C H Y K D E H E L I Y
H C E E H K U F F P M E M M
P A P P L I O R R S A V M B
M M M U D N G H E E M Y U D
Y Y D O G I E T T S H I N C
L E A S E N T Y N P P L E C
M U D S L I D E I Y H N T D
```

antibody
autoimmune
bradykinin
complement
edema
histamine
interferon
lymph
lymphokine
macrophage

a. _____ Swelling due to the accumulation of tissue fluid.
b. _____ A system of plasma proteins that are a nonspecific defense.
c. _____ A disease caused by the immune system attacking the person's own body.
d. _____ A substance found in damaged tissues that initiates nerve impulses, triggering pain.
e. _____ A molecule released by T cells that enhances the abilities of other immune system cells.

f. _____ Large phagocytic cell.
g. _____ Protein released by cells infected with viruses.
h. _____ A protein produced by B cells in response to foreign antigens.
i. _____ Substance produced by basophils or mast cells in an allergic reaction.
j. _____ Tissue fluid inside lymphatic vessels.

Chapter Test

Objective Test

Do not refer to the text when taking this test.

____ 1. The two collecting ducts of the lymphatic system empty into
   a. systemic arteries.
   b. systemic veins.
   c. pulmonary arteries.
   d. pulmonary veins.

____ 2. The structure of a lymphatic vessel is most similar to that of a
   a. cardiovascular artery.
   b. cardiovascular arteriole.
   c. cardiovascular vein.
   d. skeletal muscle fiber.

____ 3. Lymph nodes house
   a. neutrophils and monocytes.
   b. lymphocytes and macrophages.
   c. granular leukocytes.
   d. red blood cells.

____ 4. Lymph is _____ in lymphatic vessels.
   a. blood
   b. serum
   c. tissue fluid
   d. plasma

____ 5. All of the following are functions of macrophages EXCEPT
   a. liberate a growth factor to help leukocytes.
   b. phagocytize bacteria.
   c. scavenge dead and decaying tissue.
   d. transport oxygen in the blood.

Use this key for questions 6–9.
   Key *a. thymus gland   b. spleen   c. lymph node   d. red bone marrow*

____ 6. causes differentiation of T cells
____ 7. purifies lymph
____ 8. contains red pulp and white pulp
____ 9. formation of agranular and granular leukocytes

____ 10. The spleen
   a. contains stem cells from the bone marrow.
   b. is located along the trachea.
   c. produces a hormone believed to stimulate the immune system.
   d. is a blood reservoir.

____ 11. The thymus
   a. contains all types of stem cells from the bone marrow.
   b. is located along the trachea.
   c. produces a hormone believed to stimulate the immune system.
   d. both *b* and *c*

____ 12. Activity of the complement system is an example of nonspecific defense by
   a. barriers to entry.
   b. phagocytic cells.
   c. protective proteins.
   d. Both *a* and *c* are correct.

____ 13. Secretions of the oil glands is an example of nonspecific defense by a(n)
   a. barrier to entry.
   b. protective protein.
   c. phagocytic cell.
   d. acidic pH.

____ 14. Interferon is produced by cells in response to the presence of
   a. chemical irritants.
   b. viruses.
   c. bacterial infection.
   d. malarial parasite in blood.

____ 15. The most active white blood cell phagocytes are
   a. neutrophils and macrophages.
   b. neutrophils and eosinophils.
   c. lymphocytes and macrophages.
   d. lymphocytes and neutrophils.

____ 16. The white blood cells that are primarily responsible for specific immunity are
   a. neutrophils.
   b. eosinophils.
   c. macrophages.
   d. lymphocytes.

___17. Which of these is *not* a valid contrast between T cells and B cells?

              T cells   **B cells**

    a. matures in the thymus–matures in bone marrow
    b. antibody-mediated immunity–cell-mediated immunity
    c. antigen must be presented–direct recognition by APC
    d. produce lymphokines–do not produce lymphokines

___18. A particular antibody can
    a. attack any type of antigen.
    b. attack only a specific type of antigen.
    c. be produced by any B lymphocyte.
    d. be produced by any T lymphocyte.

___19. The clonal selection theory refers to the
    a. presence of four different types of T lymphocytes in the blood.
    b. response of only one type of B lymphocyte to a specific antigen.
    c. occurrence of many types of plasma cells, each producing many types of antigens.

___20. The portions of an antibody molecule that pair up with the foreign antigens are the
    a. heavy chains.
    b. light chains.
    c. variable regions.
    d. constant regions.

___21. Which of these is incorrect?
    a. helper T cells–orchestrate the immune response
    b. cytotoxic T cells–stimulate B cells to produce antibodies
    c. memory T cells–long-lasting active immunity
    d. suppressor T cells–shut down the immune response

___22. A person receiving an injection of gamma globulin as a protection against hepatitis is an example of
    a. naturally acquired active immunity.
    b. naturally acquired passive immunity.
    c. artificially acquired passive immunity.
    d. artificially acquired active immunity.

___23. A person vaccinated to produce immunity to the flu is an example of
    a. naturally acquired active immunity.
    b. naturally acquired passive immunity.
    c. artificially acquired passive immunity.
    d. artificially acquired active immunity.

___24. Allergies are caused by
    a. strong toxins in the environment.
    b. autoimmune diseases.
    c. the overproduction of IgE.
    d. the receipt of IgA in breast milk.

___25. Which is *not* true of an autoimmune response?
    a. responsible for such diseases as multiple sclerosis and myasthenia gravis and perhaps type I diabetes
    b. occurs when self-antibodies attack self-tissues
    c. interferes with the transplantation of organs between one person and another
    d. All of these are correct.

THOUGHT QUESTIONS

Answer in complete sentences.

26. How do we know that one aspect of specific immunity is the ability of the body to recognize self as opposed to nonself?

27. Describe how the lymphatic system aids the activities of the integumentary system, and vice versa.

Test Results: _____ number correct ÷ 27 = _____ × 100 = _____%

Answer Key

Study Questions

1. **a.** return of excess tissue fluid to bloodstream **b.** absorption of fats at intestinal villi **c.** defense against disease 2. **a.** F **b.** F **c.** F **d.** F **e.** T **f.** T 3. **a.** F; ... begin with lymph capillaries **b.** F; ... resembles tissue fluid that has entered the lymph vessels **c.** F; The right lymphatic duct serves the right arm, the right side of the head and neck, and the right thoracic area. **d.** F; ... capillaries form lymphatic vessels first, and these merge before entering a particular lymphatic duct 4. **a.** S **b.** RBM **c.** T **d.** S **e.** T **f.** S **g.** RBM 5. **a.** 2 **b.** 1 **c.** 3 **d.** 1 **e.** 2 **f.** 2 **g.** 1 **h.** 3 **i.** 2 **j.** 1 **k.** 4
6.

Cytotoxic T Cell		B Cell	
a.	yes	b.	yes
c.	yes	d.	no
e.	yes	f.	yes
g.	yes	h.	no
i.	no	j.	yes

7. a 8. antigens selects the B cell that will clone 9. c 10. **a.** remain in body ready to produce more antibodies when needed **b.** undergo apoptosis and die off 11. **a.** antigen-binding site **b.** variable region **c.** constant region **d.** heavy chain **e.** light chain **f.** combine with antigens and mark them for destruction 12. a 13. **a.** identify cell as belonging to a individual **b.** They are antigenic in someone else's body. 14. **a.** secretes cytokines that stimulate other immune cells **b.** destroys cells that are infected with a virus **c.** undergo apoptosis and die 15. **a.** see Figure 7.10*b*, in text **b.** immunological memory **c.** memory B **d.** memory T 16. passive immunity 17. tetanus, whooping cough, diphtheria, hepatitis B, measles, mumps, rubella, polio, Haemophilis influenza (type b) 18. It is better to prevent a disease than to try to treat it with antibiotics. Resistant strains of bacteria and allergies to antibiotics are two side effects, and viruses do not respond to antibiotic therapy. 19. **a.** Interleukins activate and maintain killer activity of T cells. **b.** Interferon causes other cells to resist a viral infection. 20. Same-type antibodies produced by the same lymphocyte. 21. **a.** Antigen attaches to IgE antibodies on mast cells and histamine release causes allergic response. **b.** Antibodies and cytotoxic T cells attack foreign antigens. **c.** Viral infection tricks immune cells into attacking tissues of self 22. d, a, h, j, e, b, i, c, g, f

Definitions Wordsearch

```
    E N I M A T S I H
    G       C O M P L E M E N T
E A N T I B O D Y D       A
N H         R     N E     U
I P         A     O M     T
K O         D     R A     O
O R         Y     E       I
H C     H K       F       M
P A   P   I       R       M
M M M     N       E       U
Y Y       I       T       N
L         N       N       E
                  I
```

a. edema **b.** complement **c.** autoimmune **d.** bradykinin **e.** lymphokine **f.** macrophage **g.** interferon **h.** antibody **i.** histamine **j.** lymph

Chapter Test

1. b 2. c 3. b 4. c 5. d 6. a 7. c 8. b 9. d 10. d 11. d 12. c 13. a 14. b 15. a 16. d 17. b 18. b 19. b 20. c 21. b 22. c 23. d 24. c 25. c 26. Ordinarily, antibodies and T cells attack only foreign antigens. If and when they attack the body's own cells, illness results. 27. The lymphatic system drains excess tissue fluid from the skin and protects against infections. The skin serves as a barrier to entry of pathogens.

8
RESPIRATORY SYSTEM

STUDY TIPS

Breathing is the first step of getting oxygen into the body so that it can eventually be used by your cells in the manufacture of ATP. In this chapter, you will learn how respiration is controlled by your brain (p. 172) and about the organs involved in this life-giving function (pp. 166–169).

Remember that the term *respiration* in biology means two different things. Cellular respiration is the metabolic pathway undertaken by your cells to break down glucose and transfer energy to ATP molecules. In this chapter, you will see that respiration also refers to the mechanisms of breathing.

When you study the material in this chapter, be sure to differentiate between external and internal respiration (p. 174). External respiration is the exchange of gases between the air inside the lungs and the blood, and internal respiration is exchange occurring between the blood and tissue cells. Make a detailed, step-by-step summary of the ways in which hemoglobin participates in these two processes (p. 174).

How environmental pollutants, including cigarette smoking (p. 181), and diseases (pp. 177–180) damage your lungs are also topics discussed here. Once you realize how a thin alveolar membrane facilitates rapid gas exchange, it will become clear why alveoli are so easily damaged.

STUDY QUESTIONS

Study the text section by section as you answer the questions that follow.

8.1 RESPIRATORY TRACT (PAGES 166–169)

- Describe the pathway air takes in and out of the lungs and the structures involved that are designed to filter, warm, and moisten air.

1. Complete this table. Refer to Table 8.1 in the textbook as needed.

Structure	Function
a. *Nasal cavities*	Filter, warm, and moisten air
Glottis	b. *passage of air to larynx*
c. *larynx*	Sound production
Trachea	d. *passage of air to bronchi*
e. *bronchi*	Passage of air to each lung
Bronchioles	f. *passage air to lung*
g. *to lungs*	Gas exchange

2. Label the following diagram, using the alphabetized list of terms.
   epiglottis
   glottis
   hard palate
   larynx
   nasal cavity
   soft palate
   trachea

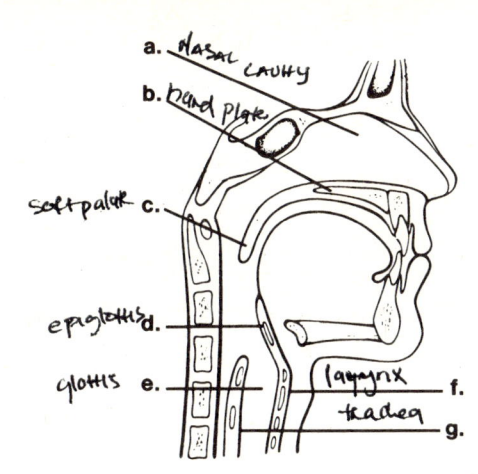

   a. nasal cavity
   b. hard palate
   c. soft palate
   d. epiglottis
   e. glottis
   f. larynx
   g. trachea

3. The nostrils contain a._____, and the nasal cavity and trachea are lined with b._____ to screen the incoming air. Mucus, dust, and other material are moved into the c._____ for swallowing or expectoration. During swallowing, the d._____ folds down over the glottis to keep food from entering the trachea. The lungs of premature infants often lack a film called e._____ that keeps their lung tissues from sticking together.

8.2 Mechanism of Breathing (pages 170–173)

- State and define the four processes involved in respiration.
- Describe the mechanism by which breathing occurs, including inspiration and expiration.

4. Label the following diagram, using the alphabetized list of terms.
   expiratory reserve volume
   inspiratory reserve volume
   residual volume
   tidal volume

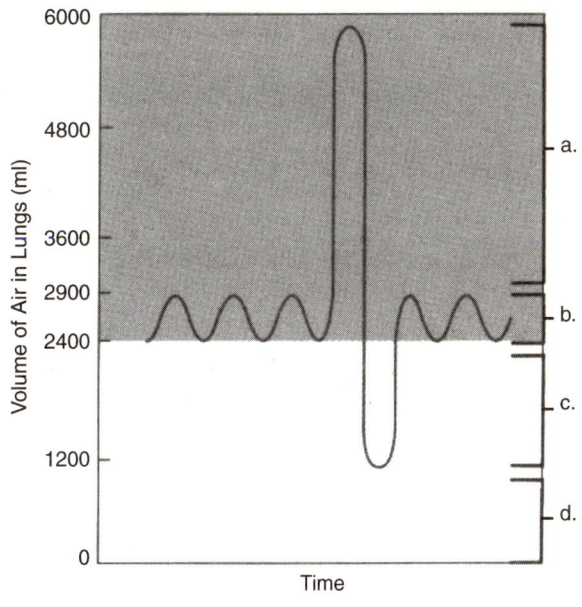

5. In the preceding diagram, the sum of the volumes labeled *a, b,* and *c* is termed the _____.

6. Indicate whether the following phrases describe inspiration (I) or expiration (E).
   __I__ a. lungs expanded
   __E__ b. muscles (diaphragm and ribs) relaxed
   __E__ c. diaphragm dome shaped
   __I__ d. chest enlarged
   __I__ e. less air pressure in lungs than in the environment

65

7. Put these statements in the proper sequence.
   _____ a. Respiratory center stops sending nerve impulse to diaphragm and rib cage.
   _____ b. Respiratory center sends nerve impulse to diaphragm and rib cage.
   _____ c. Diaphragm relaxes, becomes dome shaped, and rib cage moves down and inward.
   _____ d. Lungs expand as diaphragm lowers and rib cage moves upward and outward.
   _____ e. Air goes rushing out as lungs recoil.
   _____ f. Air comes rushing in as lungs expand.

8.3 GAS EXCHANGES IN THE BODY (PAGES 174–176)

- Describe the events that occur during external and internal respiration.
- Show that hemoglobin is well suited to its role as a respiratory pigment.

8. Match the statements with the terms below.
   _____ a. entrance and exit of air into and out of lungs
   _____ b. exchange of gases between blood and tissue fluid
   _____ c. production of ATP in cells
   _____ d. exchange of gases between lungs and blood

   Terms 1. internal respiration   2. cellular respiration   3. inspiration and expiration   4. external respiration

   e. Next, place the terms in the proper sequence.

   First _____ Second _____ Third _____ Last _____

9. Give the equation that describes how oxygen is transported in the blood. Label one arrow *lungs* and the reverse arrow *tissues*.

10. Give the equation that describes how most of the carbon dioxide is transported in the blood. Label one arrow *lungs* and the reverse arrow *tissues*.
    a.

    What is the name of the enzyme that speeds up this reaction? b. _____

    Carbon dioxide transport produces hydrogen ions. Why does the blood not become acidic? c. _____

    By what process does carbon dioxide move from the blood to the alveoli? d. _____

11. After studying Figure 8.9 in the text, answer these questions. Where does oxygen enter the blood?
    a. _____ Where does oxygen exit from the blood? b. _____
    Where does carbon dioxide enter the blood? c. _____ Where does carbon dioxide exit
    from the blood? d. _____ In the diagram, what two types of vessels are high in oxygen?
    e. _____

    In the diagram, what two types of vessels are high in carbon dioxide? f. _____

12. Hemoglobin is remarkably suited to the transport of oxygen. Why? a. _____

    Why does a person die from carbon monoxide poisoning? b. _____

    How does hemoglobin help with the transport of carbon dioxide? c. _____

8.4 RESPIRATION AND HEALTH (PAGES 177–180)

- List the names, symptoms, and causes of various diseases of the respiratory tract.
- Explain why women now have an incidence of lung cancer equal to that of men.

13. Match the descriptions in *a–g* to this key.

    Key *cold and flu   strep throat   pneumonia   tuberculosis   emphysema   pulmonary fibrosis   lung cancer*

    Cells build a protective capsule around the bacteria. X rays can detect the presence of these capsules.
    a._____ Rheumatic fever can be a complication. b._____
    Fibrous connective tissue builds up in the lungs after inhaling particles. c._____ This
    is caused by a virus, and antibiotic therapy is not helpful. d._____ This most often begins in
    a bronchus and is caused by smoking cigarettes. e._____ Lungs balloon because
    air is trapped in the alveoli. f._____ Lobules of lungs fill with fluid, caused by a
    pathogen. g._____

14. Why do women now suffer from lung cancer rates equivalent to those of men when in the past they did not?
    _____

8.5 HOMEOSTASIS (PAGE 183)

- Discuss how the respiratory system works with other systems of the body to maintain homeostasis.

15. The *Working Together* diagram in your textbook indicates ways in which the respiratory system benefits other organ systems of the body. Match the organ systems with the correct descriptions.
    a. Gas exchange in lungs provides oxygen and rids body of carbon dioxide.
    b. Gas exchange in lungs provides oxygen to skin and rids body of carbon dioxide from skin.
    c. Gas exchange increases during sexual activity.
    d. Tonsils and adenoids occur along respiratory tract; breathing aids lymph flow; lungs carry out gas exchange.
    e. Lungs excrete carbon dioxide, provide oxygen, and convert angiotensin I to angiotensin II, leading to kidney regulation.
    f. Gas exchange in lungs provides oxygen to the digestive tract and excretes carbon dioxide from the digestive tract.
    g. Lungs provide oxygen for, and rid the body of carbon dioxide from, contracting muscles.
    h. Gas exchange in lungs provides oxygen and rids body of carbon dioxide.
    i. Lungs provide oxygen for neurons and rid the body of carbon dioxide produced by neurons.
    j. Gas exchange in lungs rids body of carbon dioxide, helping to regulate the pH of blood; breathing aids venous return.

    _____ integumentary system
    _____ lymphatic system
    _____ circulatory system
    _____ muscular system
    _____ skeletal system
    _____ nervous system
    _____ urinary system
    _____ reproductive system
    _____ endocrine system
    _____ digestive system

DEFINITIONS WORDSEARCH

Review key terms by completing this wordsearch, using the following alphabetized list of terms:

```
A G L O T T I S A G A I
E X I A L V E O L U S N
P P T O R S T O P T O D
Y G I T Q Y B E G U N I
T L D G R E N L I V I A
I D A S L C I X I S N P
C A L V W O J H R O O H
A D V I I L T C I L I R
P G O D L L A T O X T A
A A L A V A A K I E A G
C V V G R L E I V S R M
L D M O I S V E M Y I J
A G E T A Y E R L B P V
T B N J J N C M P O X M
I E I M E N H A I R E I
V O C A L C O R D O O N
```

alveolus
diaphragm
epiglottis
expiration
glottis
larynx
tidal volume
ventilation
vital capacity
vocal cords

a. _____ Layer of muscle separating thoracic and abdominal cavities.
b. _____ Contains the vocal cords.
c. _____ Opening for airflow into the larynx.
d. _____ Act of expelling air.
e. _____ Amount of air involved in normal inhale/exhale cycle.
f. _____ The process of breathing.
g. _____ Fold of tissue in larynx that creates sounds.
h. _____ Structure that covers glottis during swallowing.
i. _____ Air sac in the lung.
j. _____ Maximum amount of air moved in or out of lungs during breathing.

CHAPTER TEST

OBJECTIVE TEST

Do not refer to the text when taking this test.

1. Why is oxygen needed by the body?
   a. to aerate the lungs
   b. to cleanse the blood
   c. to produce ATP
   d. Both *a* and *b* are correct.

2. The structure(s) that receive(s) air after the trachea is(are) the
   a. pharynx.
   b. bronchi.
   c. bronchiolus.
   d. alveoli.

3. Which structure carries both air and food?
   a. larynx
   b. pharynx
   c. trachea
   d. esophagus

4. How are foreign particles trapped before entering the lungs?
   a. coarse hair inside nostrils
   b. mucus in the nasal cavity
   c. cilia in the nasal cavity and trachea
   d. All of these are correct.

5. Which of these constricts during an asthma attack?
   a. trachea
   b. bronchus
   c. bronchiole
   d. pharynx

6. The alveoli
   a. are sacs in the lungs.
   b. contain capillaries.
   c. are where gas exchange occurs.
   d. All of these are correct.

__d__ 7. Which of these contains the vocal cords?
   a. glottis
   b. epiglottis
   c. pharynx
   d. larynx

__a__ 8. Before oxygen is picked up in the lungs by hemoglobin, it first diffuses through (a) alveolar cells, (b) blood plasma, (c) red blood cell plasma membranes, and (d) capillary walls, though not necessarily in this order. What is the correct order?
   a. a, b, d, c
   b. a, d, b, c
   c. d, a, c, b
   d. d, b, a, c
   e. a, b, c, d

____ 9. When the lungs recoil,
   a. inspiration occurs.
   b. external respiration occurs.
   c. internal respiration occurs.
   d. expiration occurs.
   e. All of these are correct.

____ 10. The respiratory center
   a. is stimulated by carbon dioxide.
   b. is located in the chest.
   c. sends nerve impulses to lung tissue.
   d. is stimulated by oxygen levels.

__a__ 11. The amount of air that enters or leaves the lungs during a normal respiratory cycle is the
   a. tidal volume.
   b. respiratory volume.
   c. residual volume.
   d. vital capacity.

__c b__ 12. The maximum amount of air a person can exhale after taking the deepest breath possible is a measure of the
   a. residual volume.
   b. tidal volume.
   c. vital capacity.
   d. inspiratory reserve volume.

__a__ 13. External respiration is defined as
   a. an exchange of gases in the lungs.
   b. breathing.
   c. an exchange of gases in the tissues.
   d. cellular respiration.

__b__ 14. Which gas is carried partially by the plasma?
   a. O_2
   b. CO_2
   c. both O_2 and CO_2
   d. neither O_2 nor CO_2

__b__ 15. CO_2 enters the blood as a result of
   a. active transport.
   b. diffusion.
   c. blood pressure.
   d. air pressure.

____ 16. The enzyme carbonic anhydrase causes
   a. carbon dioxide to react with water.
   b. carbon dioxide to react with bicarbonate ions.
   c. water to react with hydrogen ions.
   d. Both *b* and *c* are correct.

____ 17. Hemoglobin combines with
   a. oxygen more readily in the lungs.
   b. carbon dioxide more readily in the tissues.
   c. oxygen more readily in the tissues.
   d. carbon dioxide more readily in the lungs.
   e. Both *a* and *b* are correct.

__d__ 18. Hemoglobin carries
   a. O_2.
   b. CO_2.
   c. hydrogen ions.
   d. All of these are correct.

____ 19. Which lung disorder is *not* caused by pathogen?
   a. pneumonia
   b. tuberculosis
   c. emphysema
   d. flu

__a__ 20. Smoking cigarettes
   a. causes tuberculosis.
   b. leads to emphysema and cancer.
   c. increases the vital capacity of the lungs.
   d. leads to good health and longer life.

Match the items in the key to phrases 21–23.
   Key *a.* pneumonia *b.* lung cancer *c.* infant respiratory distress

____ 21. nonfunctional tissues interfere with gas exchange
____ 22. fluid-filled lobules
____ 23. alveolar collapse due to high surface tension
____ 24. Which body system does the respiratory system aid by providing oxygen so neurons can function properly?
   a. nervous system
   b. lymphatic system
   c. circulatory system
   d. integumentary system

____ 25. Which body system helps the respiratory system by protecting the lungs and providing points for breathing muscle attachment?
   a. circulatory system
   b. skeletal system
   c. muscular system
   d. urinary system

THOUGHT QUESTIONS

Answer in complete sentences.

26. Explain how expiration occurs once the lungs have filled with air.

27. Relate the large surface area provided by the alveoli to the process by which external respiration occurs.

Test Results: _____ number correct ÷ 27 = _____ × 100 = _____%

ANSWER KEY

STUDY QUESTIONS

1. a. nasal cavities **b.** passage of air into larynx **c.** larynx **d.** passage of air to bronchi **e.** bronchi **f.** passage of air to lungs **g.** lungs **2. a.** nasal cavity **b.** hard palate **c.** soft palate **d.** epiglottis **e.** glottis **f.** larynx **g.** trachea **3. a.** coarse hairs **b.** cilia **c.** pharynx **d.** epiglottis **e.** surfactant **4. a.** inspiratory reserve volume **b.** tidal volume **c.** expiratory reserve volume **d.** residual volume **5.** vital capacity **6. a.** I **b.** E **c.** E **d.** I **e.** I **7.** b, d, f, a, c, e **8. a.** 3 **b.** 1 **c.** 2 **d.** 4 **e.** 3, 4, 1, 2

9. $Hb + O_2 \underset{tissues}{\overset{lungs}{\rightleftharpoons}} HbO_2$

10. a. $CO_2 + H_2O \underset{lungs}{\overset{tissues}{\rightleftharpoons}} H_2CO_3 \underset{lungs}{\overset{tissues}{\rightleftharpoons}} H^+ + HCO_3^-$

b. carbonic anhydrase **c.** Hemoglobin combines with excess hydrogen ions. **d.** diffusion **11. a.** lungs **b.** tissues **c.** tissues **d.** lungs **e.** pulmonary vein and aorta (systemic arteries) **f.** venae cavae (systemic veins) and pulmonary artery **12. a.** It easily combines with oxygen in the lungs and easily gives it up in the tissues. **b.** Hemoglobin combines with carbon monoxide preferentially to oxygen. **c.** It combines with carbon dioxide to a degree, called carbaminohemoglobin, and picks up hydrogen ions from the equation of 10a. **13. a.** tuberculosis **b.** strep throat **c.** pulmonary fibrosis **d.** cold and flu **e.** lung cancer **f.** emphysema **g.** pneumonia **14.** Women now smoke cigarettes as frequently as men. In the past, it was less acceptable for women to smoke. **15.** b, d, j, g, a, i, e, c, h, f

DEFINITIONS WORDSEARCH

```
        G L O T T I S
    E     A L V E O L U S
    P T   R           D
Y   I     Y           I
T   D G       N       A
I   A   L       X   N P
C   L     O         O O H
A   V       T     I   I R
P   O         T       T A
A   L           A I   A G
C   V             L   S R M
L   M I               I
A   E T               P
T   N                 X
I   E                 E
V O C A L C O R D
```

a. diaphragm **b.** larynx **c.** glottis **d.** expiration **e.** tidal volume **f.** ventilation **g.** vocal cords **h.** epiglottis **i.** alveolus **j.** vital capacity

CHAPTER TEST

1. c **2.** b **3.** b **4.** d **5.** c **6.** d **7.** d **8.** b **9.** d **10.** a **11.** a **12.** c **13.** a **14.** b **15.** b **16.** b **17.** e **18.** d **19.** c **20.** b **21.** b **22.** a **23.** c **24.** a **25.** b **26.** The diaphragm relaxes when the respiratory center stops sending messages to contract. Once relaxation occurs, expiration is passive. Air leaves the lungs with the elastic recoil of the lungs. Muscle contraction can force additional air from the lungs. **7.** Since oxygen enters the capillaries of the alveoli by the process of diffusion, a passive process, a large surface area is required.

9
URINARY SYSTEM AND EXCRETION

STUDY TIPS

The human kidney is an intriguing structure designed to conserve water and to excrete a concentrated urine. Unless you take the time to study the topics of this chapter carefully, it is easy to become overwhelmed by the variety of events occurring in the nephron, the functional unit of the kidney.

Begin your study of this chapter by first reviewing the structures involved in the urinary system (p. 188). Learn the macroscopic structure of the kidney (p. 191), then go inside it and learn about the nephron (pp. 192–193).

Make a diagram of a nephron and label its parts. Next, using a different color of ink or pencil, write in the functions of each section of nephron. Next, tie in the process of urine formation (p. 195) with the structure of the nephron.

Now you are ready to study how different features of this system are controlled by hormones (pp. 196–197). This chapter ends by examining kidney dialysis (p. 200), the treatment used to remove metabolic wastes from the blood when the kidneys no longer function.

STUDY QUESTIONS

Study the text section by section as you answer the questions that follow.

9.1 URINARY SYSTEM (PAGES 188–189)

- List the organs involved in the urinary system.
- State, in general, the contents of urine.
- Trace the path of urine, and describe the general structure and function of each organ mentioned.
- Describe how urination is controlled by the nervous system.
- Explain how the kidneys function to maintain homeostasis of the body's internal environment.

1. Match the urinary organs listed in the key to functions *a–d*.
   Key kidney   ureter   urinary bladder   urethra

   ureter a. muscular tube leading from kidneys to urinary bladder
   urethra b. tube leading from bladder to the outside
   urinary bladder c. hollow, muscular organ that stores urine
   kidney d. bean-shaped organ that filters blood

2. What triggers urination? _____

3. List five functions of the kidneys.
   a. _____
   b. _____
   c. _____
   d. _____
   e. _____

9.2 KIDNEYS (PAGES 191–193)

- Describe the macroscopic structure of the kidney.
- State the parts of a kidney nephron and relate these to the macroscopic anatomy.

4. Macroscopically, the kidney is composed of these three parts.

   a. _____

   b. _____

   c. _____

5. Label the parts of the nephron on the following diagram, using the alphabetized list of terms:

   afferent arteriole
   collecting duct
   distal convoluted tubule
   efferent arteriole
   glomerular capsule
   glomerulus
   loop of the nephron
   peritubular capillaries
   proximal convoluted tubule

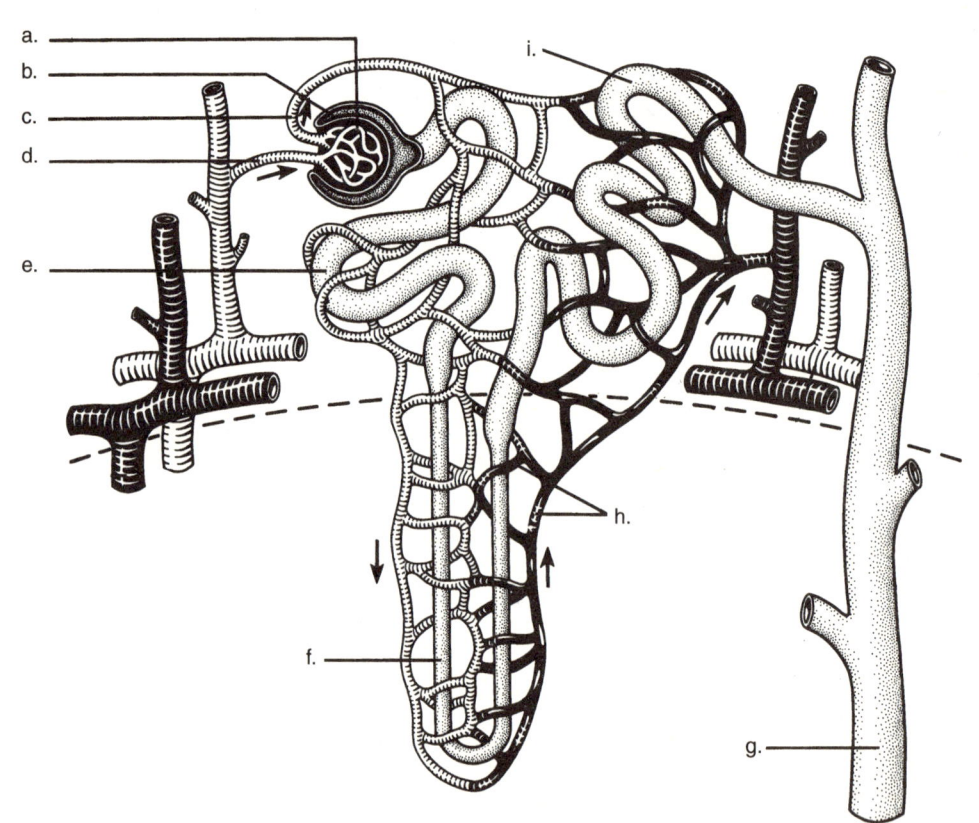

   a. _____
   b. _____
   c. _____
   d. _____
   e. _____
   f. _____
   g. _____
   h. _____
   i. _____

6. Trace the path of filtrate from the glomerular capsule to the collecting duct.
   glomerular capsule

   a. _____

   b. _____

   c. _____

   collecting duct

72

9.3 URINE FORMATION (PAGE 195)

- Describe the three steps in urine formation, and relate these to parts of a nephron.

7. In this diagram, add the three steps in urine formation.

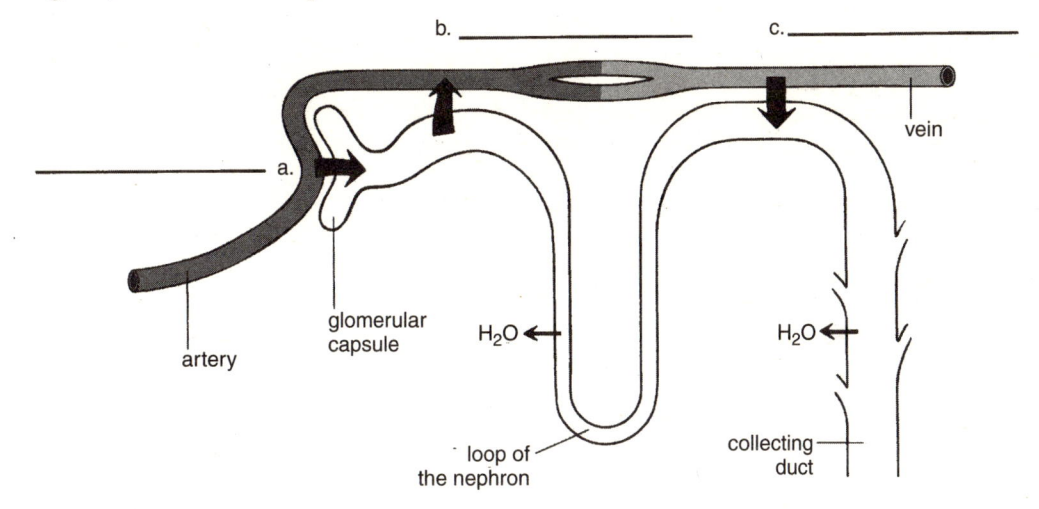

Where does the first step take place? d. _____
Where does the second step take place? e. _____
Where does the third step take place? f. _____

8. Blood in the glomerulus is composed of these two portions:

   Small Molecules   **Large Molecules, etc.**
   Nutrients (glucose, amino acids)   Proteins
   Wastes (urea, uric acid)   Formed elements
   Salts
   Water

   Which portion will undergo filtration?
   a. _____

   Which of the small molecules will maximally undergo reabsorption?
   b. _____

   Which of the small molecules will minimally undergo reabsorption?
   c. _____

9. Give an example of a molecule that undergoes tubular excretion. _____

9.4 MAINTAINING WATER-SALT BALANCE (PAGES 196–197)

- Describe how the loop of the nephron contributes to water reabsorption.
- Name three hormones involved in maintaining blood volume, and explain how they function.

10. Water reabsorption. The presence of which parts of the nephron accounts for maximal reabsorption of water in humans? a. _____ What causes water to leave these parts of the nephron?
    b. _____

    Because water is maximally reabsorbed, humans excrete a c. _____ urine.

11. Complete this table.

ADH	Urine Quality	
Increased amount	a.	
Reduced amount	b.	

12. Match the items in the key to the following letters.

    Key *blood pressure rises, adrenal cortex, converting enzyme, renin, aldosterone, atrial natriuretic hormone*

    a. _____ changes angiotensinogen to angiotensin I

    b. _____ changes angiotensin I to angiotensin II

    Angiotensin II acts on c. _____ to secrete d. _____

    Kidneys absorb Na+ and e. _____

    When blood pressure rises, the heart secretes f. _____, which causes the kidneys to excrete Na+.

9.5 Maintaing Acid-Based Balance (page 199)

- Describe how the kidneys excrete hydrogen ions and reabsorb bicarbonate ions to regulate the pH of blood.

13. If the blood is acidic, a. _____ ions are excreted in combination with b. _____, while c. _____ are reabsorbed. If the blood is basic, fewer d. _____ ions are excreted, and fewer e. _____ and are reabsorbed.

9.6 Homeostasis (page 199)

- Explain how the activities of the urinary system and other body systems work together to maintain homeostasis.

14. The *Working Together* diagram in your textbook indicates ways in which the urinary system benefits other organ systems of the body. Match the organ systems with the correct descriptions.
    a. Kidneys control volume of body fluids, including lymph.
    b. Kidneys maintain blood levels of Na+, K+, and Ca++, which are needed for muscle innervation, and eliminate creatinine, a muscle waste.
    c. Kidneys compensate for water loss through respiratory tract; work with lungs to maintain blood pH.
    d. Kidneys provide active vitamin D for Ca++ absorption and help maintain blood level of Ca++, needed for bone growth and repair.
    e. Kidneys convert vitamin D to active form needed for Ca++ absorption; compensate for any water loss by digestive tract.
    f. Kidneys maintain blood levels of Na+, K+, and Ca++, which are needed for nerve connection.
    g. Kidneys keep blood values within normal limits so that transport of hormones continues.
    h. Kidneys filter blood and excrete wastes; maintain blood volume, pressure, and pH; produce renin and erythropoietin.
    i. Semen is discharged through the urethra in males; kidneys excrete wastes and maintain electrolyte levels for mother and child.
    j. Kidneys compensate for water loss due to sweating; activate vitamin D precursor made by skin.

    _____ integumentary system   _____ nervous system
    _____ lymphatic system   _____ urinary system
    _____ circulatory system   _____ reproductive system
    _____ muscular system   _____ endocrine system
    _____ skeletal system   _____ digestive system

9.7 PROBLEMS WITH KIDNEY FUNCTION (PAGE 200)

• Describe how the kidney machine works.

15. Both of the following indicate a problem with kidney function. Why?

    albumin in the urine a. _____

    high blood urea b. _____

16. a. Place arrows for urea in this simplified diagram illustrating a hemodialysis machine.

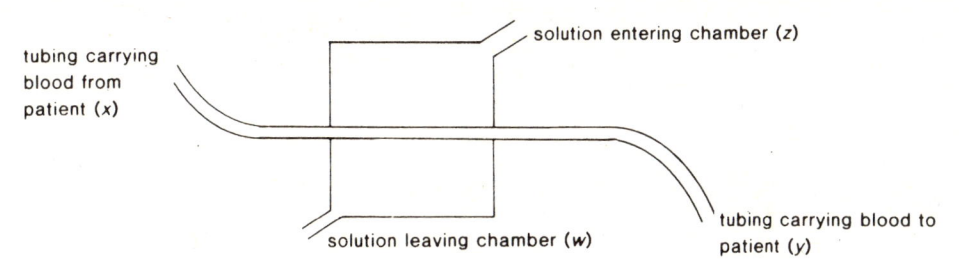

    tubing carrying blood from patient (x)
    solution entering chamber (z)
    solution leaving chamber (w)
    tubing carrying blood to patient (y)

    b. In general, what should be the makeup of the solution that enters the chamber at z? _____
    _____

Excretion Elimination

In the table, place an X beside the component of blood if the following descriptions pertain to it:

  a. in the afferent arteriole
  b. in the filtrate
  c. in the efferent arteriole
  d. reabsorbed into the peritubular capillary
  e. secreted from the peritubular capillary
  f. present in urine
  g. absent from urine
  h. in venous blood

	a	b	c	d	e	f	g	h
1. plasma proteins								
2. red blood cells								
3. white blood cells								
4. glucose								
5. amino acids								
6. sodium chloride								
7. water								
8. urea								
9. uric acid								
10. penicillin								

There are 45 correct answers. There are 80 possible errors of omission or commission. Any 10 errors and you're ELIMINATED!

DEFINITIONS CROSSWORD

Review key terms by completing this crossword puzzle, using the following alphabetized list of terms:

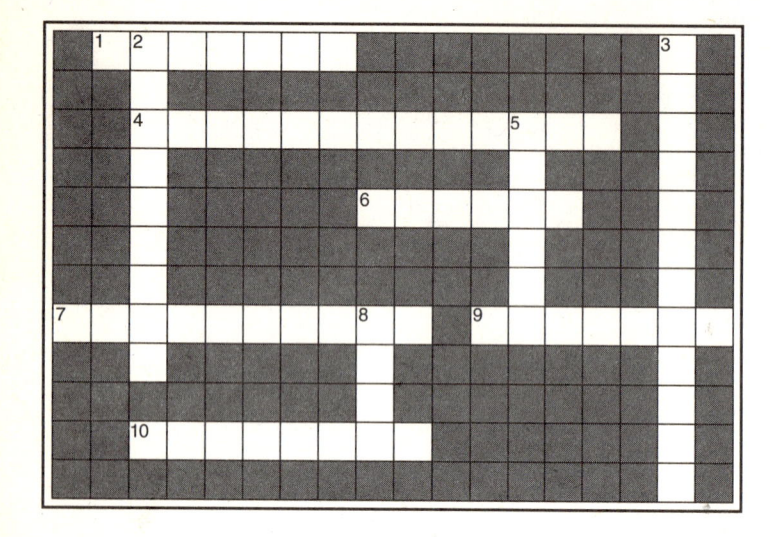

antidiuretic
collecting duct
excretion
glomerulus
kidney
nephron
proximal
urea
ureter
urethra

Across
1. Anatomical and functional unit of the kidney.
4. Tube that receives urine from the distal convoluted tubules of several nephrons.
6. Organ of the urinary system that produces and excretes urine.
7. The ball of capillaries, surrounded by a glomerular capsule, in the nephron.
9. Tube conveying urine from the bladder to outside the body.
10. Portion of convoluted tubule nearest the glomerular capsule.

Down
2. Process of removing metabolic wastes from the body.
3. Type of hormone from the posterior pituitary that promotes the reabsorption of water from the collecting duct.
5. One of two tubes leading from the kidneys to the urinary bladder.
8. Main nitrogenous waste derived from the breakdown of amino acids.

CHAPTER TEST

OBJECTIVE TEST

Do not refer to the text when taking this test.

_e_ 1. Kidneys are organs of homeostasis because they
   a. regulate the blood volume.
   b. regulate the pH of the blood.
   c. help maintain the correct concentration of ions in the blood.
   d. excrete nitrogenous wastes.
   e. All of these are correct.

_a_ 2. Which of these contains urine?
   a. urethra
   b. uterus
   c. intestine
   d. gallbladder
   e. All of these are correct.

_d_ 3. Which portion of the urinary tract varies significantly in length or size between males and females?
   a. kidneys
   b. ureters
   c. urinary bladder
   d. urethra

_a_ 4. Urination is triggered by
   a. contraction of the bladder and relaxation of sphincter muscles.
   b. relaxation of the bladder and contraction of sphincter muscles.
   c. contraction of kidney and ureter muscles.
   d. contraction of kidneys and relaxation of the urinary bladder.

___ 5. Which portion is *not* part of the kidney?
   a. renal cortex
   b. renal urethra
   c. renal medulla
   d. renal pelvis

_b_ 6. The collecting ducts are primarily in the
   a. renal cortex.
   b. renal medulla.
   c. renal pelvis.
   d. afferent arteriole.

Match the function in 7–11 with the structure in the key.
___ 7. extends into the medulla
___ 8. a knot of capillaries
___ 9. variably permeable to water
___ 10. site of afferent/efferent arterioles
___ 11. blind end of the proximal convoluted tubule
   Key a. glomerulus b. glomerular capsule
      c. renal cortex d. loop of the nephron
      e. collecting duct

_c_ 12. Urine collects in the _____
   before entering the ureter.
   a. renal medulla
   b. renal cortex
   c. renal pelvis
   d. capsule

_d_ 13. Glomerular filtration should be associated with
   a. the glomerular capsule.
   b. the distal convoluted tubule.
   c. the collecting duct.
   d. All of these are correct.

___ 14. Sodium is removed from the kidney tubule by
   a. passive reabsorption.
   b. active reabsorption.
   c. an attraction to Cl⁻.
   d. secretion.

___ 15. Tubular secretion occurs at
   a. the glomerular capsule.
   b. the proximal convoluted tubule.
   c. the loop of the nephron.
   d. the distal convoluted tubule.

___ 16. In humans, water is
   a. found in the glomerular filtrate.
   b. reabsorbed from the nephron.
   c. in the urine.
   d. All of these are correct.

___ 17. Glucose
   a. is in the filtrate and urine.
   b. is in the filtrate and not in urine.
   c. undergoes tubular secretion and is in urine.
   d. undergoes tubular secretion and is not in urine.

___ 18. The loop of the nephron is characteristic of animals that excrete
   a. a diluted urine.
   b. a concentrated urine.
   c. no urine.
   d. too much urine.

___ 19. Aldosterone
   a. is secreted by the adrenal cortex.
   b. causes the blood volume to lower.
   c. is the same as renin.
   d. causes the kidneys to excrete sodium.

___ 20. Which organ excretes bile pigments?
   a. kidney
   b. large intestine
   c. liver
   d. All of these are correct.

___ 21. ADH is necessary for
   a. water reabsorption.
   b. glucose reabsorption.
   c. protein reabsorption.
   d. All of these are correct.

___ 22. If the nephrons do not function,
   a. urea accumulates in the blood.
   b. edema occurs.
   c. water and salt retention occur.
   d. All of these are correct.

___ 23. Renin is an enzyme that converts
   a. angiotensinogen to angiotensin I.
   b. angiotensin I to angiotensin II.
   c. angiotensin II to converting enzyme.
   d. All of these are correct.

___ 24. Which body system removes excess tissue fluid, thus maintaining blood pressure and ensuring proper kidney function?
   a. respiratory system
   b. endocrine system
   c. lymphatic system
   d. reproductive system

___ 25. The kidneys provide active vitamin D for calcium absorption that directly benefits which body system?
   a. skeletal system
   b. circulatory system
   c. digestive system
   d. integumentary system

Thought Questions

Answer in complete sentences.

26. Explain the difference between defecation and excretion.

27. What role is played by the high concentration of salt and urea in the renal medulla?

Test Results: _____ number correct ÷ 27 = _____ × 100 = _____ %

Answer Key

Study Questions

1. a. ureter **b.** urethra **c.** urinary bladder **d.** kidney
2. Stretch receptors send impulses to the spinal cord, which sends nerve impulses back to muscles controlling the urinary bladder. Contraction of the bladder and relaxation of sphincter muscles occur, expelling urine to the outside. **3. a.** regulation of salt and water balance **b.** regulation of blood volume and blood pressure **c.** stimulation of red blood cell production **d.** regulation of the pH of the blood **e.** removal of metabolic wastes, including nitrogenous wastes **4. a.** renal cortex **b.** renal medulla **c.** renal pelvis **5. a.** glomerulus **b.** glomerular capsule **c.** efferent arteriole **d.** afferent arteriole **e.** proximal convoluted tubule **f.** loop of the nephron **g.** collecting duct **h.** peritubular capillaries **i.** distal convoluted tubule
6. a. proximal convoluted tubule **b.** loop of the nephron **c.** distal convoluted tubule **7. a.** glomerular filtration **b.** tubular reabsorption **c.** tubular secretion **d.** between the glomerulus and the glomerular capsule **e.** between the proximal convoluted tubule and blood **f.** between blood and the distal convoluted tubule **8. a.** small molecules **b.** nutrients, salts, water **c.** wastes **9.** uric acid, hydrogen ions, ammonia, penicillin, creatinine **10. a.** loop of the nephron and the collecting duct **b.** hypertonic medulla due to presence of salt (from ascending limb of loop of the nephron) and urea (from collecting duct) **c.** hypertonic **11. a.** little urine **b.** much urine **12. a.** renin **b.** converting enzyme **c.** adrenal cortex **d.** aldosterone **e.** blood pressure rises **f.** atrial natriuretic hormone **13. a.** hydrogen **b.** ammonia **c.** bicarbonate ions **d.** hydrogen **e.** bicarbonate ions **14.** j, a, h, b, d, f, c, i, g, e **15. a.** Normally, albumin does not filter out of the glomerulus. Increased permeability indicates kidney failure. **b.** Normally, urea filters out of the glomerulus. **16. a.** Arrows should go from the tubing to w. **b.** Solution should contain those substances that are not to leave the blood and should lack those substances that are to leave the blood.

Game: Excretion Elimination

	a.	b.	c.	d.	e.	f.	g.	h.
plasma proteins	X		X				X	X
red blood cells	X		X				X	X
white blood cells	X		X				X	X
glucose	X	X		X			X	X
amino acids	X	X		X			X	X
sodium chloride	X	X		X		X		X
water	X	X	X	X		X		X
urea	X	X				X		X
uric acid	X	X			X	X		
penicillin	X		X		X	X		

Definitions Crossword

Across
1. nephron 4. collecting duct 6. kidney 7. glomerulus 9. urethra 10. proximal

Down
2. excretion 3. antidiuretic 5. ureter 8. urea

Chapter Test

1. e **2.** a **3.** d **4.** a **5.** b **6.** b **7.** d **8.** a **9.** e **10.** c **11.** b **12.** c **13.** a **14.** b **15.** d **16.** d **17.** b **18.** b **19.** a **20.** c **21.** a **22.** d **23.** a **24.** c **25.** a **26.** Defecation is the elimination of nondigested material from the gut, and excretion is the elimination of end products of metabolism by the kidneys. **27.** The high concentration of salt and urea in the renal medulla draws water out of the loop of the nephron and the collecting duct.

PART 3 MOVEMENT AND SUPPORT IN HUMANS

10

SKELETAL SYSTEM

STUDY TIPS

The skeletal system has the job of protecting and supporting the organs of the body. It also provides points of attachment for muscles and contributes to homeostasis by storing minerals and producing blood cells for the body.

Begin your study of the skeletal system by learning about the structure of bone (pp. 206–207) and the types of tissues involved (p. 206). Sketch your own diagram of a long bone (see p. 207), and label functions associated with different portions.

Bone formation and remodeling are continuous throughout a lifetime. Make a chart with two columns in which you can compare the two types of bone formation, intramembranous ossification (p. 208) and endochondral ossification (p. 208). Also make lists of the steps involved in bone remodeling (p. 209) and in bone repair (p. 209).

The best way to learn the names of the skeletal bones (pp. 211–217) is to study them repeatedly and often. Refer to a clear diagram that labels bones, and quiz yourself as you learn them. Pronounce their names out loud. Saying and hearing a name while reading it will reinforce it in your memory. Review the types of joints (p. 218) and joint movements (p. 219). Finish your study of this chapter by using the *Working Together* illustrations (p. 220) to review how the skeleton interacts with other systems of the body.

STUDY QUESTIONS

Study the text section by section as you answer the questions that follow.

10.1 TISSUES OF THE SKELETAL SYSTEM (PAGE 206–207)
- Describe the bone, cartilage, and fibrous connective tissues that comprise the skeleton.
- Sketch and label the structure of a long bone.

1. Compact bone is composed of bone cells arranged in concentric circles called a._____. Inside compact bone, a lighter type of bone, called b._____, has spaces that are often filled with c._____, the blood-forming tissue. Ligaments, which attach bone to bone, are made up of d._____ tissue.

2. Name the three types of cartilage and where they are found.

   a._____
   b._____
   c._____

3. Label the following diagram of a long bone, using the alphabetized list of terms.
   cartilage
   compact bone
   medullary cavity
   spongy bone

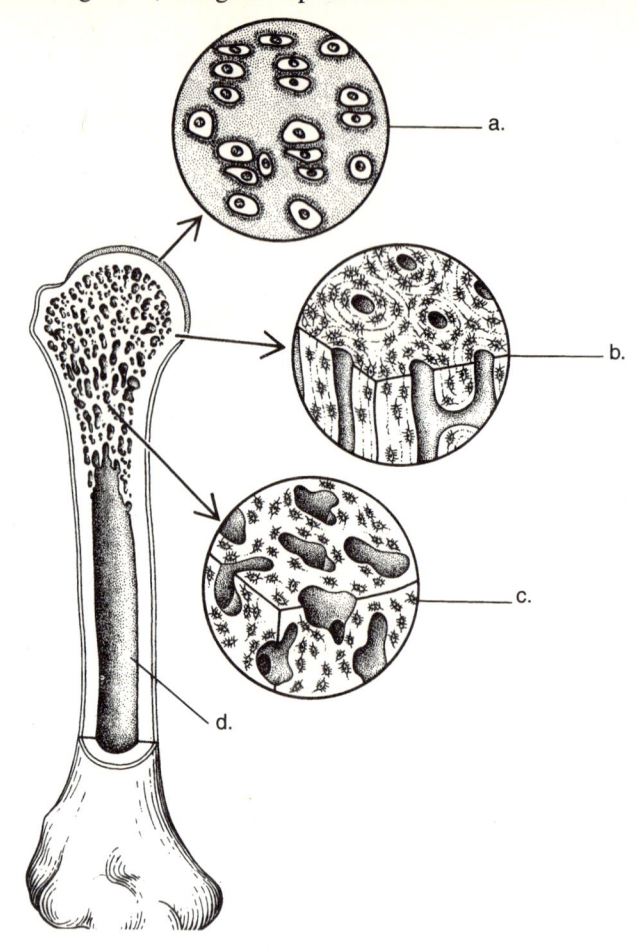

   a. _____
   b. _____
   c. _____
   d. _____

4. Which part of the long bone shown in the previous illustration is associated with red bone marrow?
   a. _____ Which part of bone is the hardest? b. _____
   Which part of bone is the most flexible? c. _____ Which part of a long bone is associated with yellow bone marrow? d. _____

10.2 BONE GROWTH AND REPAIR (PAGES 208–209)

- List the different cell types of bone, and give their functions.
- Differentiate between intramembranous and endochondral ossification.
- Understand the function of the epiphyseal plates of developing long bones.
- Explain the process of remodeling of bones.
- List the sequence of events that occur as a bone repairs a fracture.

5. What is the relationship between osteoblasts and osteocytes? a. _____
   What role is played by the cartilaginous disks of long bones? b. _____
   What is the function of an osteoclast? c. _____
   What is the process of bone formation called? d. _____
   What are the cells that are unspecialized precursors to other bone cells? e. _____

6. Place an *I* for intramembranous ossification or an *E* for endochondral ossification in the spaces that follow.
   _____ a. skull bones
   _____ b. cartilaginous models of bones
   _____ c. epiphyseal plates remain as the bones grow in length
   _____ d. spongy bone laid down first, compact bone forms over top
   _____ e. bones develop between sheets of fibrous connective tissue

7. During the continual process of remodeling of bone, osteoclasts derived from a._____ in red bone marrow remove worn-out bone cells. At the same time, b._____ is released into the bloodstream. The two factors that affect bone thickness are c._____ and d._____.

8. Repair of fractures. Place the following steps in the proper sequence by rearranging the letters.
   _d_ a. fibrocartilage callus ✓
   _c_ b. bony callus ✓
   _b_ c. remodeling ✓
   _a_ d. hematoma forms ✓

10.3 BONES OF THE SKELETON (PAGES 211–217)

- List the functions of the skeleton.
- Identify and state a function for the bones of the axial skeleton, including the cranium and face.
- Identify and state a function for the bones of the appendicular skeleton.

9. Name five functions of the skeleton.
   a. _protects_
   b. _stores calcium/phosphorus_
   c. _attachment of muscles_
   d. _movmt of body_
   e. _formation of cells_

10. Axial versus appendicular skeleton:
    Write *ax* in front of all bones belonging to the axial skeleton; write *ap* in front of all bones belonging to the appendicular skeleton. Write *pec* in front of all bones belonging to the pectoral girdle; write *pel* in front of all bones belonging to the pelvic girdle.
    _ap_ a. coxal bone   _ax_ g. ribs
    _ax_ b. sternum   _pel_ h. radius
    pel c. humerus   _pec_ i. clavicle
    _ap_ d. scapula   _pel_ j. tibia
    _ax_ e. skull   _pel_ k. fibula
    pel f. femur   _pel_ l. ulna

11. The a._____ bone forms the forehead, and the b._____ bones has the foramen magnum through which the spinal cord passes. The c._____ bones have an opening for the ears. The d._____ is the only movable portion of the skull and permits us to chew our food. The e._____ forms the upper jaw and anterior hard plate. The f._____ bones form the cheekbone, and the nasal bones form the bridge of the nose.

12. Give a function of vertebrae. a._____
    Give the name of the vertebrae that have ribs. b._____
    Give the name of the vertebrae in the lower back. c._____
    Give the name of the vertebrae in the neck. d._____

13. Comparisons. The radius and ulna are to the lower arm as the a._____ and b._____ are to the lower leg. The femur is to the upper leg as the c._____ is to the upper arm. The metacarpals are to the palm as the d._____ are to the foot.

14. Give the scientific terms for the common names.
    Shinbone a._____
    Collarbone b._____
    Hipbone c._____
    Thighbone d._____

81

10.4 ARTICULATIONS (PAGES 218–219)

- Classify joints according to their types.
- List the different types of joint movements.

15. Match the type of joint listed in the key with its description.
    Key *fibrous joint   cartilaginous joint   synovial joint*

    Synovial joint a. lined with synovial membrane
    fibrous b. immovable, as in a suture
    synovial c. slightly movable
    fibrous joint d. freely movable
    cartilaginous e. connected by hyaline cartilage

16. Fill in the table to describe the shoulder and elbow joints.

Joint	Anatomical Type	Degree of Movement
Shoulder joint	a.	b.
Elbow joint	c.	d.

17. Label the following diagram to indicate the various types of joint movements, using the alphabetized list of terms.
    abduction, adduction, eversion, extension, flexion, inversion, rotation, supination/pronation

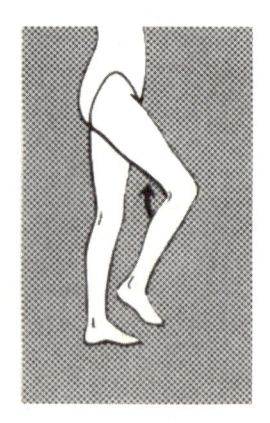

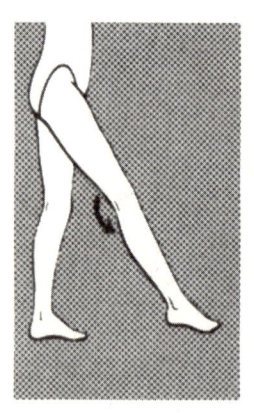

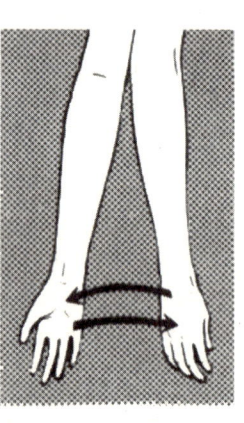

    a. _____   b. _____   c. _____   d. _____

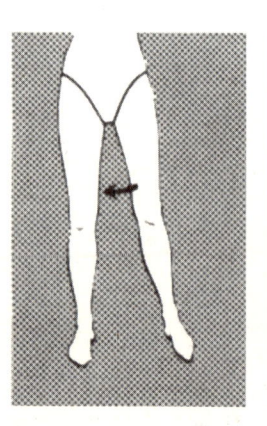

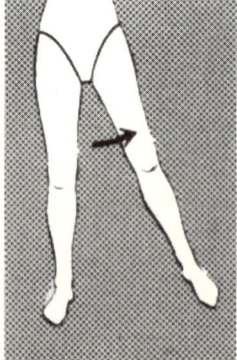

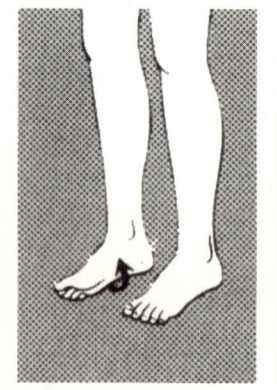

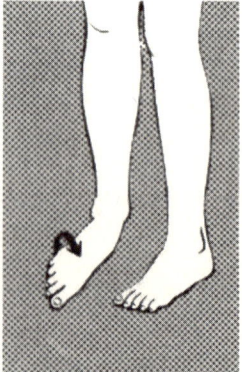

    e. _____   f. _____   g. _____   h. _____

82

10.5 Homeostasis (page 221)

- Discuss how the skeletal system works with other systems of the body to maintain homeostasis.

18. The *Working Together* diagram in your textbook shows how the skeletal system benefits from the activities of the other organ systems of the body. Match the organ systems with the correct descriptions.
    a. Muscular contraction causes bones to move joints; muscles help protect bones.
    b. Gas exchange in lungs provides oxygen and rids body of carbon dioxide.
    c. Kidneys provide active vitamin D for Ca^{++} absorption and help maintain blood level of Ca^{++}, needed for bone growth and repair.
    d. Growth hormone regulates bone development; parathyroid hormone and calcitonin regulate Ca^{++} content.
    e. Digestive tract provides Ca^{++} and other nutrients for bone growth and repair.
    f. Blood vessels deliver nutrients and oxygen to bones, carry away wastes.
    g. Sex hormones influence bone growth and density in males and females.
    h. Receptors send sensory input from bones to joints.
    i. Skin protects bones; helps provide vitamin D for Ca^{++} absorption.
    j. Lymphatic vessels pick up excess tissue fluid; immune system protects against infections.

    _____ integumentary system
    _____ lymphatic system
    __f__ circulatory system
    __a__ muscular system
    _____ urinary system
    _____ nervous system
    __b__ respiratory system
    _____ reproductive system
    _____ endocrine system
    __e__ digestive system

Definitions Wordsearch

Review key terms by completing this wordsearch, using the following alphabetized list of terms:

```
O O S U F E M A E C A R T O P E R
E T J C O M P A C T A C I L B R Y
R O M M N A R I Y L A J O I N T N
C F U F T A R E T A L B L G N T E
N I E B A L O R S I I S S A S S A
U T T L N E T U S X U M G M O O B
O E S M E E T T S A L C O E T S O
A R O E L L L U Y O U A B N N R I
S T I O U T J S E S U S L T J U G
B A R E D B O N E M A R R O W M R
N A E M T A E R G C R R E F I N E
A P P E N D I C U L A R O V E A D
```

appendicular
axial
compact
fontanel
ligament
joint
osteoclast
periosteum
red bone marrow
suture

a. _Osteoclast_ Cell that breaks down old bone.
b. _red bone marrow_ Blood-forming tissue housed inside spongy bone.
c. _appendicular_ Portion of the skeleton made up of arms and legs.
d. _axial_ Portion of skeleton along midline.
e. _Joint_ An articulation.
f. _fontanel_ Soft spot in newborn skull.
g. _compact_ Hardest portion of a long bone.
h. _suture_ Skull joint.
i. _periosteum_ Fibrous connective tissue covering each bone.
j. _ligament_ Fibrous connective tissue band binding bone to bone.

Chapter Test

Objective Test

Do not refer to the text when taking this test.

_b_ 1. The hardest portion of a long bone is/are the
 a. articular cartilages.
 b. compact bone.
 c. spongy bone.
 d. ligaments.

_a_ 2. Bone-building cells are _____, while mature bone cells are called _____.
 a. osteoblasts; osteocytes
 b. osteoclasts; osteoblasts
 c. osteoclasts; osteocytes
 d. osteoblasts; lacunae

_d_ 3. Where is the blood-forming tissue housed within a bone?
 a. throughout the bone
 b. along the periosteum
 c. within the compact bone
 d. within the spongy bone

_d_ 4. Which of these is a function of the skeleton?
 a. support
 b. protection
 c. production of red blood cells
 d. all of these

_a_ 5. Ligaments join
 a. bone to bone.
 b. muscle to muscle.
 c. muscle to bone.
 d. All of these are correct.

_d_ 6. Endochondral ossification involves
 a. long bones.
 b. a cartilaginous model.
 c. an epiphyseal plate.
 d. All of these are correct.

84

__d__ 7. Where in the body does intramembranous ossification occur?
   a. tibia
   b. humerus
   c. femur
   d. skull bones

__c__ 8. Choose the proper sequence describing repair of bone fractures.
   a. remodeling–fibrocartilage callus–hematoma–bony callus
   b. fibrocartilage callus–hematoma–remodeling–bony callus
   c. hematoma–fibrocartilage callus–bony callus–remodeling
   d. bony callus–remodeling–hematoma–fibrocartilage callus

__d__ 9. Where do the cells that remodel bone originate?
   a. osteocytes that revert to earlier stages
   b. red blood cells that undergo morphogenesis
   c. from osteoblasts in yellow bone marrow
   d. from monocytes in red bone marrow

__b__ 10. Which type of fracture is usually due to twisting?
   a. incomplete fracture
   b. spiral fracture
   c. comminuted fracture
   d. impacted fracture

__d__ 11. Which of these is *not* a function of the skeleton?
   a. supports the body
   b. protects soft body parts
   c. stores calcium and phosphorus
   d. conduction of nervous impulses

__d__ 12. Which of these is *not* in the appendicular skeleton?
   a. clavicle
   b. coxal bone
   c. metatarsals
   d. vertebrae

__a__ 13. Which of these is a facial bone?
   a. frontal bone
   b. occipital bone
   c. mandible
   d. All of these are correct.

__c__ 14. Mastoiditis, a complication of severe ear infections, is most apt to occur in which portion of the skull?
   a. fontanels
   b. frontal bone
   c. mastoid sinuses
   d. parietal bones

__b__ 15. Which bone articulates with all others of the cranium and is thus considered the keystone bone?
   a. ethmoid
   b. sphenoid
   c. temporal
   d. frontal

__a__ 16. Which of these is *not* a bone in the leg?
   a. femur
   b. tibia
   c. ulna
   d. fibula

__c__ 17. Vertebrae have
   a. immovable joints.
   b. freely movable joints.
   c. slightly movable joints.
   d. joints that vary from one person to the next.

__d__ 18. The shoulder joint is an example of a
   a. synovial joint.
   b. freely movable joint.
   c. ball-and-socket joint.
   d. All of these are correct.

__c__ 19. The bones making up your palm are the _____.
   a. phalanges
   b. carpals
   c. metacarpals
   d. metatarsals

__b__ 20. Which vertebrae support the lower back?
   a. cervical
   b. lumbar
   c. thoracic
   d. cranial

__c__ 21. Which type of joint is most likely to develop arthritis?
   a. fibrous
   b. cartilaginous
   c. synovial
   d. epiphyseal disk

__a__ 22. Fluid-filled sacs within joints are called
   a. bursae.
   b. menisci.
   c. articulations.
   d. synovial fluid.

__a__ 23. Bending your knee so that your heel touched your buttock would be called
   a. flexion.
   b. extension.
   c. eversion.
   d. pronation.

__c__ 24. The system that provides calcium and other nutrients for bone growth and repair of the skeleton is the _____ system.
   a. urinary
   b. digestive
   c. endocrine
   d. lymphatic

__d__ 25. The system that has receptors to send information to and from bones and joints is the _____ system.
   a. respiratory
   b. integumentary
   c. nervous
   d. circulatory

THOUGHT QUESTIONS

Answer in complete sentences.

26. Why are knee injuries usually quite serious?

27. Describe the remodeling of bone.

Test Results: _____ number correct ÷ 27 = _____ × 100 = _____%

ANSWER KEY

STUDY QUESTIONS

1. a. osteons **b.** spongy bone **c.** red bone marrow **d.** fibrous connective **2. a.** hyaline cartilage; ends of bones **b.** fibrocartilage; intervertebral disks **c.** elastic cartilage; ear flaps **3. a.** cartilage **b.** compact bone **c.** spongy bone **d.** medullary cavity **4. a.** spongy bone **b.** compact bone **c.** cartilage **d.** medullary cavity **5. a.** bone-forming osteoblasts eventually become mature osteocytes **b.** they increase in length and allow bone to grow longer **c.** to break down bone **d.** ossification **e.** osteogenitor cells **6. a.** I **b.** E **c.** E **d.** I **e.** I **7. a.** monocytes **b.** calcium **c.** exercise **d.** hormones **8.** d, a, b, c **9. a.** supports the body **b.** protects soft body parts **c.** produces blood cells **d.** stores calcium and phosphorus salts **e.** permits body movement **10. a.** ap, pel **b.** ax **c.** ap **d.** ap, pec **e.** ax **f.** ap **g.** ax **h.** ap **i.** ap, pec **j.** ap **k.** ap **l.** ap **11. a.** frontal **b.** occipital **c.** temporal **d.** mandible **e.** maxilla **f.** zygomatic **12. a.** to protect the spinal cord **b.** thoracic **c.** lumbar **d.** cervical **13. a.** tibia **b.** fibula **c.** humerus **d.** metatarsals **14. a.** tibia **b.** clavicle **d.** coxal bone **e.** femur **15. a.** synovial **b.** fibrous **c.** cartilaginous **d.** synovial **e.** cartilaginous **16. a.** ball-and-socket (synovial) **b.** freely movable **c.** hinge (synovial) **d.** freely mov-able **17. a.** flexion **b.** extension **c.** rotation **d.** supina-tion (top), pronation (bottom) **e.** adduction **f.** abduction **g.** inversion **h.** eversion **18.** i, j, f, a, c, h, b, g, d, e

DEFINITIONS WORDSEARCH

```
            F
       C O M P A C T       L
M N           L   J O I N T
U T       E   A       G
E A       R   I       A
T N       U   X       M
S E       T S A L C O E T S O
O L       U           N
I         S           T
R E D B O N E M A R R O W
E
A P P E N D I C U L A R
```

a. osteoclast **b.** red bone marrow **c.** appendicular **d.** axial **e.** joint **f.** fontanel **g.** compact **h.** suture **i.** periosteum **j.** ligament

CHAPTER TEST

1. b **2.** a **3.** d **4.** d **5.** a **6.** a **7.** d **8.** c **9.** d **10.** b **11.** d **12.** d **13.** c **14.** c **15.** b **16.** c **17.** c **18.** d **19.** c **20.** b **21.** c **22.** a **23.** a **24.** b **25.** c **26.** The knee is a complex, large joint with many ligaments and menisci for strength and stability. Injuring just one area of the knee means much pain and time spent waiting for healing. The knee also supports the weight of the body, which can make healing a very slow process. **27.** Monocytes from red bone marrow become osteoclasts that tear down worn-out bone tissue. During this process, calcium is released into the bloodstream. Osteo-blasts build and remodel the bone according to the needs of the body.

11
MUSCULAR SYSTEM

STUDY TIPS

The muscular system provides several functions for your body aside from movement. This chapter focuses on the anatomy and physiology of skeletal muscle.

Review the gross anatomy of a whole muscle (p. 226). Understand that one end of the muscle is called the insertion while the other is the origin (p. 227). Muscles work in groups, some together, and others in opposition (p. 227). Spend time learning about how muscles are named (p. 228), which will help you remember names of the major muscles of the body (p. 229) and their function (p. 229). Cover up muscle names in Fig. 11.4 and quiz yourself.

Within muscle cells there are contractile units called sarcomeres (p. 231). Study their structure, then learn how contraction occurs as filaments within sarcomeres slide past each other (p. 230).

Briefly review your notes on cellular respiration from chapter 2 before you study the section on energy for muscle contraction (p. 235). Recall the function of ATP in the cell. ATP is again employed by muscle fibers for contraction. Muscle fibers also use backup sources of energy, as you will discover as you read Section 11.4 in the text.

The chapter ends with a discussion of the need for exercise for muscle strength and health and describes two types of muscle fibers (p. 237) that are adapted to different types of exercise. Now you will be able to more fully appreciate why you may be a long-distance runner while your friend is better adapted to lifting weights!

STUDY QUESTIONS

Study the text section by section as you answer the questions that follow.

11.1 SKELETAL MUSCLES (PAGES 226–230)

- Describe the anatomy of a whole muscle.
- List the functions of skeletal muscle.
- Name the major muscles of the body and know their actions.

1. Individual muscles are wrapped in a coat of fibrous connective tissue, called a. _____. When a group of muscles work together, the one pulling the greatest load is called the b. _____, while the other helper muscles are called c. _____. Muscles that work opposite to each other are known as d. _____.

2. Muscle cells, known as a. _____, contain thin, or b. _____, filaments and thick, or c. _____, filaments. The myofilaments are arranged in contractile units called d. _____.

3. List three functions of skeletal muscle.
   a. _____
   b. _____
   c. _____

87

4. Label the following drawing of muscles and bones in the arm, using the alphabetized list of terms.
   biceps brachii
   humerus
   insertion
   origin
   radius
   scapula
   triceps brachii
   ulna

5. With reference to the muscles in the drawing for question 4, why are the biceps brachii and triceps brachii antagonistic pairs? _____
   _____

6. Name the thigh muscles that act as antagonists.
   Front of thigh a. _____
   Back of thigh b. _____

7. Name the antagonistic muscles of the lower leg.
   Front of lower leg a. _____
   Back of lower leg b. _____

11.2 MECHANISM OF MUSCLE FIBER CONTRACTION (PAGES 231–233)

- Describe the anatomy of a muscle fiber.
- Explain how the sarcomere shortens during muscle contraction.
- Describe the neuromuscular junction, and tell how impulses are transferred to the muscle sarcolemma.

8. The plasma membrane of a muscle fiber is named the a. _____. A special kind of endoplasmic reticulum, called b. _____ _____, stores the element c. _____ in the cell. d. _____ are invaginations of the plasma membrane.

9. Label the following drawing of a portion of a muscle fiber, using the alphabetized list of terms.
   myofibrils
   openings to T tubules
   sarcolemma

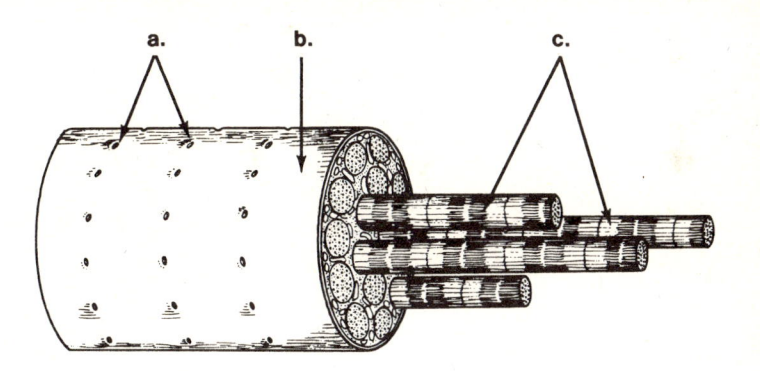

   a. _____ b. _____ c. _____

10. Label the following drawing of a sarcomere, using the alphabetized list of terms.
    actin filament
    H zone
    myosin filament
    Z line

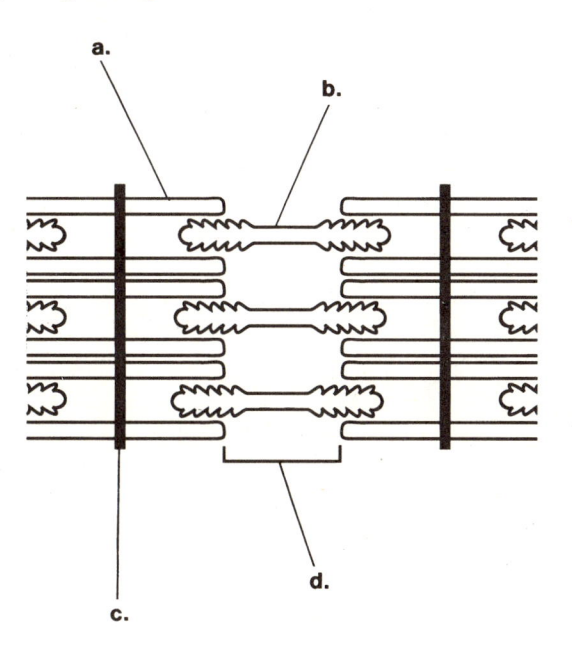

11. Which of your labels in question 10 is a thin filament? a. _____ Which of your labels is a thick filament? b. _____ Which of your labels is reduced in size when a sarcomere contracts? c. _____ Which component has cross-bridges? d. _____ Which of your labels is the filament that moves when the sarcomere contracts? e. _____ What molecule immediately supplies energy for muscle contraction? f. _____

12. Put the following phrases in the proper order, by rearranging the letters, to describe what occurs at the neuromuscular junction to trigger muscle contraction.
    ____ a. receptor sites on sarcolemma
    ____ b. nerve impulse
    ____ c. release of calcium from sarcoplasmic reticulum
    ____ d. the neurotransmitter acetylcholine is released
    ____ e. sarcomeres shorten
    ____ f. synaptic cleft
    ____ g. spread of impulses over sarcolemma to T tubules

13. Study this diagram and fill in the blanks following.

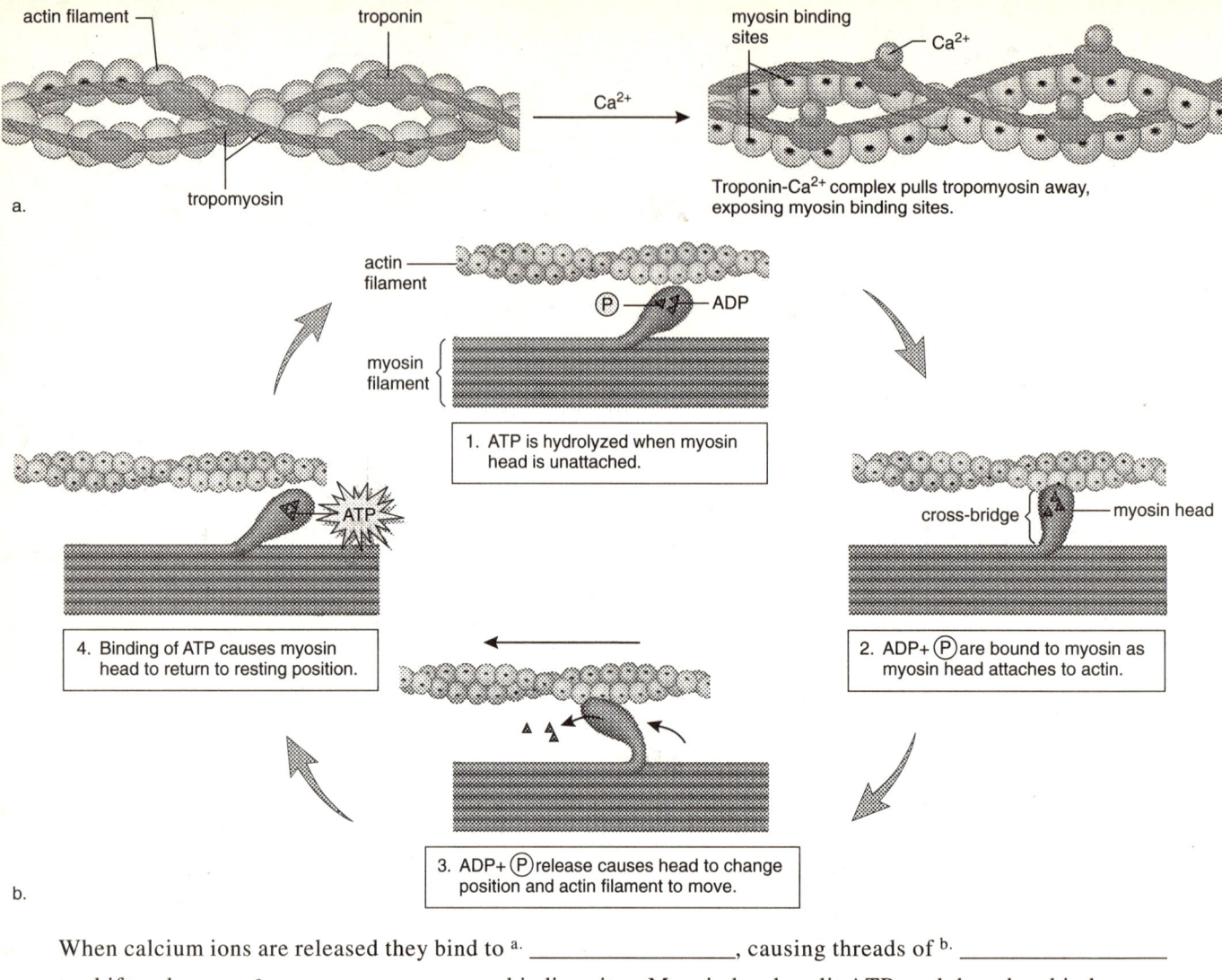

a.

b.

When calcium ions are released they bind to a. _____, causing threads of b. _____ to shift and expose c. _____ binding sites. Myosin heads split ATP, and then they bind to d. _____, forming a e. _____. The release of ADP + P from the myosin heads causes the cross-bridges to f. _____ their position and g. _____ the actin filaments toward the middle of the sarcomere.

11.3 WHOLE MUSCLE CONTRACTION (PAGE 234)

- Describe the basic laboratory experiments on whole muscle contraction.
- Explain how muscle tone is maintained.

14. Label this diagram of a single muscle twitch, using the alphabetized list of terms.
    contraction
    latent period
    relaxation

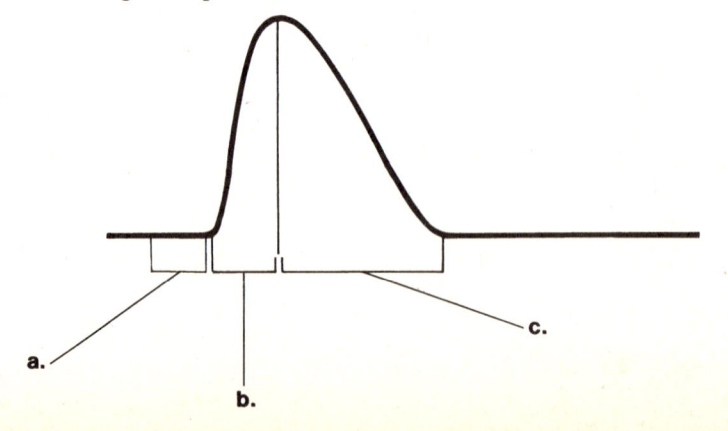

15. Indicate the locations on this myogram where the stimulus was applied, where fatigue begins, where tetanus occurs, and where the time interval is shown.

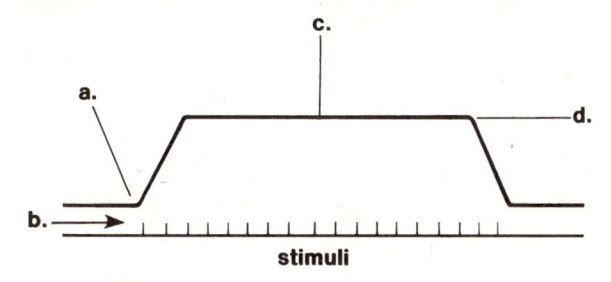

16. How do you recognize fatigue of a muscle? _____
17. Muscle tone is achieved because of special nerve receptors called _____ that send the brain information about when the muscle needs to contract.
18. Stimulation of a muscle fiber results in a contraction that can be described as _____.

11.4 ENERGY FOR MUSCLE CONTRACTION (PAGES 235–237)

- List the sources of energy available within a muscle fiber, and explain the circumstances in which each is used.
- Discuss the events leading to oxygen debt.

19. Compare sources of energy and oxygen for muscle contraction. Use the terms in the key to match with these statements.
    Key ATP   creatine phosphate   myoglobin   fermentation

    _____ a. anaerobic source of energy
    _____ b. produced during aerobic respiration
    _____ c. used to regenerate ATP from ADP
    _____ d. stores oxygen in muscle fibers
    _____ e. results in the buildup of lactic acid

20. Oxygen debt occurs when a._____ is used up in muscles, and the blood does not supply b._____ rapidly enough. Why are marathon runners less apt to become exhausted due to oxygen debt? c._____

21. Muscle wasting is called a._____, while muscle building is called b._____. To build muscles, one must lift weights or exercise to achieve what percent of maximum muscle contraction? c._____ Muscles increase in size because there is an increase in the number of d._____.

22. Use the terms in the key to match with the following statements.
    Key slow-twitch fibers   fast-twitch fibers

    _____ a. more common in marathon runners
    _____ b. more common in weight lifters
    _____ c. fewer mitochondria
    _____ d. white in color
    _____ e. abundant reserve of fat and glycogen
    _____ f. can contract for longer periods
    _____ g. contract with more force

91

11.5 HOMEOSTASIS (PAGE 238)

- Discuss how the muscular system works with other systems of the body to maintain homeostasis.

23. The *Working Together* diagram in your textbook shows how the muscular system works with the other organ systems of the body. Match the organ systems with the correct descriptions.

    a. *Muscle contractions keep blood moving in heart and blood vessels.*
    b. *Muscle contraction causes bones to move joints; muscles help protect bones.*
    c. *Muscle contraction provides heat to warm skin.*
    d. *Skeletal muscle contraction moves lymph; physical exercise enhances immunity.*
    e. *Muscles help protect glands.*
    f. *Smooth muscle contraction accounts for peristalsis; skeletal muscles support and help protect abdominal organs.*
    g. *Muscle contraction assists breathing; physical exercise increases respiratory capacity.*
    h. *Muscle contraction occurs during orgasm and moves gametes; abdominal and uterine muscle contraction occurs during childbirth.*
    i. *Smooth muscle contraction assists voiding of urine; skeletal muscles support and help protect urinary organs.*
    j. *Muscle contraction moves eyes, permits speech, creates facial expressions.*

    _____ integumentary system
    _____ lymphatic system
    _____ circulatory system
    _____ skeletal system
    _____ urinary system
    _____ nervous system
    _____ respiratory system
    _____ reproductive system
    _____ endocrine system
    _____ digestive system

DEFINITIONS WORDSEARCH

Review key terms by completing this wordsearch, using the following alphabetized list of terms:

```
T E L L E Y I N I T C A
E D N S O R R Y T I Y M
T A O I A M H I E S O M
A T D Y B P A B T O N E
N H N M O R I G I N P L
U E E R O L M E T O L O
S M T E V I H T I W O C
S A R C O M E R E P P R
B M Y O S I N S T E I A
Y N O I T R E S N I N S
```

actin
atrophy
insertion
myosin
origin
sarcolemma
sarcomere
tendon
tetanus
tone

a. _____ Structural and functional unit of a myofibril.
b. _____ Wasting of muscle due to lack of use.
c. _____ End of muscle attached to the movable bone.
d. _____ End of muscle attached to immovable bone.
e. _____ Plasma membrane of a muscle cell.
f. _____ Tension that is still present in resting muscle.
g. _____ Protein of thin filaments.
h. _____ Protein of thick filaments.
i. _____ Joins muscle to bone.
j. _____ Sustained maximal muscle contraction.

Chapter Test

Objective Test

Do not refer to the text when taking this test.

____ 1. The fascia surrounding a muscle is made up of
   a. cartilage.
   b. fibrous connective tissue.
   c. blood vessels.
   d. adipose tissue.

____ 2. The contractile unit of a skeletal muscle is the
   a. myofibril.
   b. actin filament.
   c. sarcomere.
   d. myosin filament.

____ 3. A muscle pulling just opposite to another can be best described as a(n)
   a. protagonist.
   b. antagonist.
   c. prime mover.
   d. synergist.

____ 4. Which of these is a function of skeletal muscle?
   a. produce heat
   b. posture
   c. movement
   d. All of these are correct.

____ 5. Choose the muscle that is the antagonist to the triceps brachii.
   a. gastrocnemius
   b. quadriceps femoris
   c. biceps brachii
   d. latissimus dorsi

____ 6. Choose the muscle that is the antagonist to the tibialis anterior.
   a. gastrocnemius
   b. sartorius
   c. gluteus maximus
   d. trapezius

____ 7. Which muscle covers the shoulder and functions to raise the arm?
   a. external oblique
   b. deltoid
   c. occipitalis
   d. peroneus longus

____ 8. Which muscle is located on the front thigh and raises (extends) the lower leg?
   a. trapezius
   b. rectus abdominis
   c. pectoralis major
   d. quadriceps femoris

____ 9. Which portion of the muscle is on the stationary bone?
   a. the insertion
   b. the origin
   c. the bursa
   d. the belly

____ 10. Which of these is the smallest unit?
   a. muscle fiber
   b. myofibril
   c. sarcomere
   d. actin

____ 11. When sarcomeres contract, they get shorter, and this requires that muscles work in antagonistic pairs.
   a. true
   b. false

____ 12. Myosin is
   a. the thick filament of a sarcomere.
   b. a protein.
   c. an organelle.
   d. Both *b* and *c* are correct.

____ 13. Which of these contains the cross-bridges?
   a. the actin
   b. the myosin
   c. the Z line
   d. All of these are correct.

____ 14. According to the sliding filament theory,
   a. actin moves past myosin.
   b. myosin moves past actin.
   c. both myosin and actin move past each other.
   d. None of these is correct.

____ 15. The release of calcium from the sarcoplasmic reticulum
   a. causes the sarcomeres to relax.
   b. causes sarcomeres to contract.
   c. is the end result of a nervous impulse to contract.
   d. Both *b* and *c* are correct.
   e. All of these are correct.

____ 16. When a nervous impulse reaches the synaptic end bulb,
   a. neurotransmitter is released into the synaptic cleft.
   b. calcium is released from the sarcoplasmic reticulum.
   c. calcium is stored in the T tubules.
   d. neurotransmitter is released by the sarcolemma.

____ 17. The junction between nerve and muscle is
   a. called the neuromuscular junction.
   b. where a neurotransmitter travels from the nerve fiber to the muscle fiber.
   c. the region where nerves innervate muscles.
   d. All of these are correct.

____ 18. The Z line of a sarcomere is where the
   a. cross-bridges attach.
   b. actin filaments attach.
   c. myosin filaments attach.
   d. nerve innervates the muscle cell.

93

___19. Muscle fatigue
   a. follows summation and tetanus.
   b. involves the buildup of lactic acid.
   c. occurs only in the laboratory.
   d. Both *a* and *b* are correct.
   e. All of these are correct.
___20. Oxygen debt may be associated with
   a. anaerobic cellular respiration.
   b. fermentation.
   c. muscle contraction.
   d. lactic acid metabolism.
   e. All of these are correct.
___21. Creatine phosphate is
   a. used by sarcomeres.
   b. used to change ADP to ATP.
   c. a molecule found in DNA.
   d. All of these are correct.
___22. Hypertrophy occurs only when
   a. lactic acid accumulates.
   b. athletes take steroids.
   c. muscles contract to at least 75% of maximum tension.
   d. obesity abates.

___23. Fast-twitch fibers
   a. are white in color.
   b. are associated with strength training.
   c. tend to be anaerobic.
   d. All of these are correct.
___24. Which system provides oxygen and removes carbon dioxide from muscles?
   a. integumentary system
   b. skeletal system
   c. respiratory system
   d. urinary system
___25. The muscular system aids the flow of materials through vessels of which system?
   a. lymphatic system
   b. circulatory system
   c. urinary system
   d. All of these are correct.

Thought Questions

Answer in complete sentences.
26. Why does muscle contraction depend on the presence of myofibrils in a muscle fiber?

27. A nervous action potential eventually causes a neurotransmitter to be secreted at a synapse. What event occurs as a result of a muscle action potential?

Test Results: _____ number correct ÷ 27 = _____ × 100 = _____ %

Answer Key

Study Questions

1. a. fascia **b.** prime mover **c.** synergists **d.** antagonists
2. a. fibers **b.** actin **c.** myosin **d.** sarcomeres   **3. a.** provides posture **b.** movement **c.** provides heat   **4. a.** scapula **b.** origin **c.** humerus **d.** biceps brachii **e.** triceps brachii **f.** radius **g.** insertion **h.** ulna   **5.** The biceps brachii raises and the triceps brachii lowers the forearm.   **6. a.** quadriceps femoris group **b.** hamstring muscles   **7. a.** tibialis anterior **b.** gastrocnemius   **8. a.** sarcolemma **b.** sarcoplasmic reticulum **c.** calcium **d.** T tubules   **9. a.** openings to T tubules **b.** sarcolemma **c.** myofibrils   **10. a.** actin filament **b.** myosin filament **c.** Z line **d.** H zone
11. a. actin filament **b.** myosin filament **c.** H zone **d.** myosin **e.** actin filament **f.** ATP   **12.** b, d, f, a, g, c, e
13. a. troponin **b.** tropomyosin **c.** myosin **d.** actin **e.** cross-bridge **f.** change **g.** pull   **14. a.** latent period **b.** contraction **c.** relaxation   **15. a.** stimulus applied **b.** time interval **c.** tetanus occurs **d.** fatigue begins
16. The muscle relaxes even though a stimulus has been applied.   **17.** muscle spindles   **18.** all-or-none   **19. a.** fermentation, creatine phosphate **b.** ATP **c.** creatine phosphate **d.** myoglobin **e.** fermentation   **20. a.** ATP **b.** oxygen **c.** Marathon runners have more mitochondria.
21. a. atrophy **b.** hypertrophy **c.** 75% **d.** myofibrils
22. a. slow-twitch fibers **b.** fast-twitch fibers **c.** fast-twitch fibers **d.** fast-twitch fibers **e.** slow-twitch fibers **f.** slow-twitch fibers **g.** fast-twitch fibers   **23.** c, d, a, b, i, j, g, h, e, f

Definitions Wordsearch

```
T              N I T C A
E N            Y         M
T O            H         M
A D        P   T O N E
N N   O R I G I N       L
U   E R                 O
S   T                   C
S A R C O M E R E       R
  M Y O S I N           A
  N O I T R E S N I     S
```

a. sarcomere **b.** atrophy **c.** insertion **d.** origin **e.** sarcolemma **f.** tone **g.** actin **h.** myosin **i.** tendon **j.** tetanus

Chapter Test

1. b **2.** c **3.** d **4.** d **5.** c **6.** a **7.** b **8.** d **9.** b **10.** d **11.** a **12.** d **13.** b **14.** a **15.** d **16.** a **17.** d **18.** b **19.** d **20.** e **21.** b **22.** c **23.** d **24.** c **25.** d **26.** Myofibrils contain the contractile proteins, namely myosin and actin filaments.   **27.** A muscle action potential causes calcium to be released from calcium storage sacs.

PART 4 INTEGRATION AND COORDINATION IN HUMANS

12

NERVOUS SYSTEM

STUDY TIPS

The nervous system has the ultimate control over homeostasis of the body. It is one of two principal control centers of the body, with the endocrine system (chapter 14) as the other.

In this chapter, you will learn about different types of neurons (pp. 246–247) and their structures and functions. Nervous impulse conduction (p. 248) makes use of the sodium-potassium pump, a feature of every body cell. You already know about a synapse between a neuron and a muscle fiber from the previous chapter. Synapses between neurons are no different except that a number of neurotransmitters are involved (p. 251), each producing a different effect.

By far the most complex part of the nervous system is the brain (pp. 258–261). Make a sketch of the brain, labeling its parts. Then, using a different color of ink or pencil, write in the unique functions of each area of the brain. Cover your labels and quiz yourself to help you remember the different portions of the brain.

STUDY QUESTIONS

Study the text section by section as you answer the questions that follow.

12.1 NEURONS AND HOW THEY WORK (PAGES 246–251)

- Describe the structure and function of the three major types of neurons.

1. Every neuron has the three parts listed here. What is the function of each?

   dendrite a._____

   cell body b._____

   axon c._____

2. Label the parts of the sensory neuron, the interneuron, and the motor neuron, using the alphabetized list of terms. Note: some terms may be used more than once.
   axon, axon bulb, cell body, dendrite, effector, node of Ranvier, nucleus of Schwann cell, sensory receptor

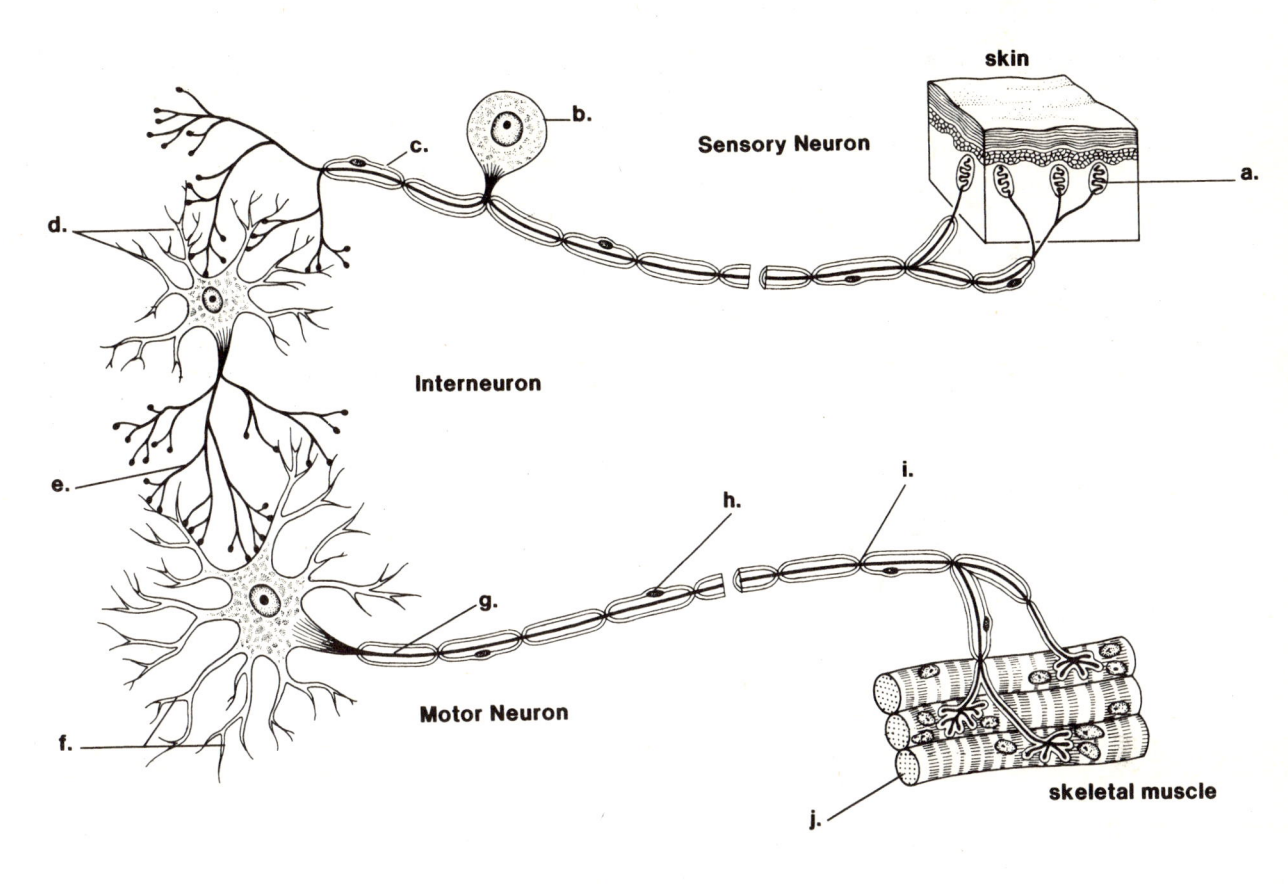

3. State the function of the complete sensory neuron. _____

4. State the function of the complete interneuron. _____

5. State the function of the complete motor neuron. _____

97

Nerve Impulse (Page 248)

- Describe the nerve impulse as an electrochemical change that can be recorded by means of an oscilloscope.

6. The following drawings represent axons. On the left, place (+) or (−) to indicate the polarity asked for. On the right, indicate the distribution of ions that produces this polarity.

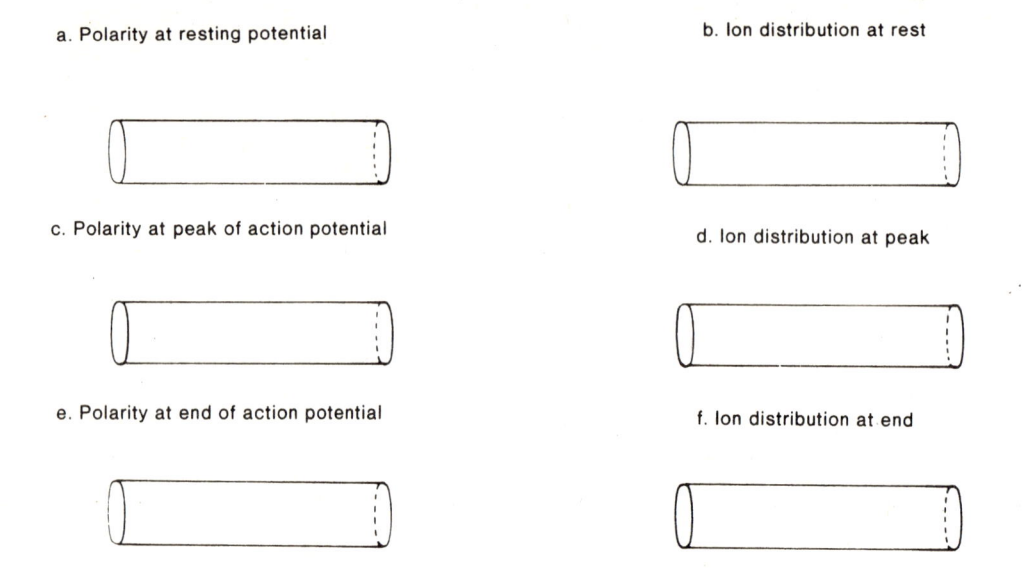

    a. Polarity at resting potential

    b. Ion distribution at rest

    c. Polarity at peak of action potential

    d. Ion distribution at peak

    e. Polarity at end of action potential

    f. Ion distribution at end

7. On this drawing of the trace that appears on the oscilloscope screen during the time of the action potential, label on one side *Na⁺* (sodium ions) *gates open* and on the other side *K⁺* (potassium ions) *gates open*. Write in the appropriate values to indicate the resting potential and the peak potential.

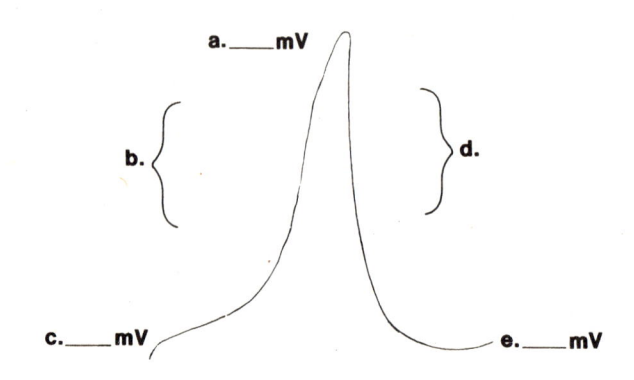

    a. ____ mV
    b. {
    c. ____ mV
    d. }
    e. ____ mV

8. During the time of rest, the _____-_____ pump restores the original distribution of ions across the membrane of a nerve fiber.

Transmission Across a Synapse (Page 251)

• Describe the structure and function of a synapse, including transmission across a synapse.

9. Label the following diagrams, using the alphabetized list of terms.
   axon, dendrite, neurotransmitter, postsynaptic membrane, synapse, synaptic cleft, synaptic vesicle

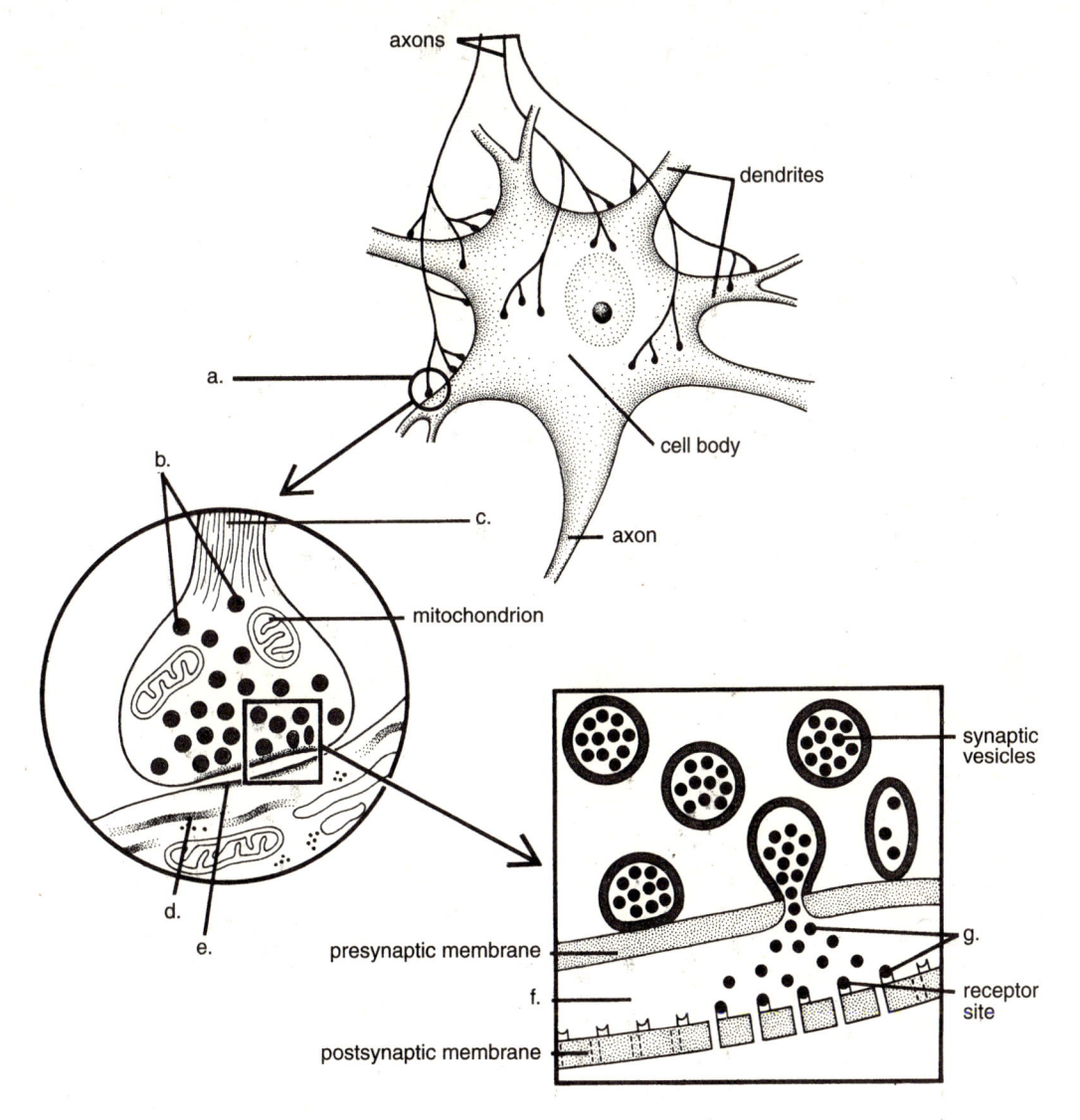

10. What causes the transmission of the nerve impulse across a synapse? _____
_____
_____

11. Which of these statements is true concerning synaptic integration?
    _____ a. A single neuron synapses with one other neuron.
    _____ b. A single neuron synapses with many other neurons.
    _____ c. Excitatory signals have a hyperpolarizing effect and inhibitory signals have a depolarizing effect.
    _____ d. Integration is the summing up of excitatory and inhibitory signals.
    _____ e. The more inhibitory signals received, the more likely the axon will conduct a nerve impulse.

12.2 PERIPHERAL NERVOUS SYSTEM (PAGES 252–255)

- Define a nerve, and distinguish between spinal and cranial nerves.
- Cite the various divisions of the nervous system and the particular functions of each division.
- Describe the path of a spinal reflex.
- Describe the autonomic system, and cite similarities as well as differences in the structure and function of the two divisions.

12. Name two types of nerves in the peripheral nervous system. a. _____ and b. _____ nerves. What is a nerve? c. _____. Why is a spinal nerve called a mixed nerve? d. _____

13. Label the following diagram of the reflex arc, using the alphabetized list of terms.
    dorsal root ganglion
    effector
    interneuron
    motor fiber
    sensory fiber
    sensory receptor

14. Fill in these blanks to indicate what happens during a spinal reflex arc.
    A stimulus is received by a(n) a. _____, which initiates an impulse in the b. _____ fiber. The sensory fiber takes the message to the spinal cord and transmits it to the c. _____. This neuron passes the impulse to the d. _____ neuron, whose fiber takes the message from the cord and innervates a muscle, causing a reaction to the stimulus.

15. Explain how the brain becomes aware of automatic reflex actions. a. _____
    _____
    _____
    Explain why the left side of the brain controls the right side of the body. b. _____
    _____
    _____

16. Indicate three ways in which the sympathetic and parasympathetic systems are similar.
    a. _____
    b. _____
    c. _____

17. Indicate ways in which the two systems differ by filling in the following table.

	Sympathetic	Parasympathetic
Type of situation	a.	b.
Neurotransmitter	c.	d.
Ganglia near cord, or Ganglia near organ?	e.	f.
Spinal nerves only, or Spinal nerves plus vagus?	g.	h.

12.3 CENTRAL NERVOUS SYSTEM (PAGES 256–259)

- State the function of neuroglorial cells.
- Describe, in general, the anatomy of the brain, name five major parts, and give a function of each.
- Name the lobes of the cerebrum, and give a function of each.

18. Label this diagram of the spinal cord, using the alphabetized list of terms.
    central canal
    dorsal horn
    gray matter
    ventral horn
    white matter

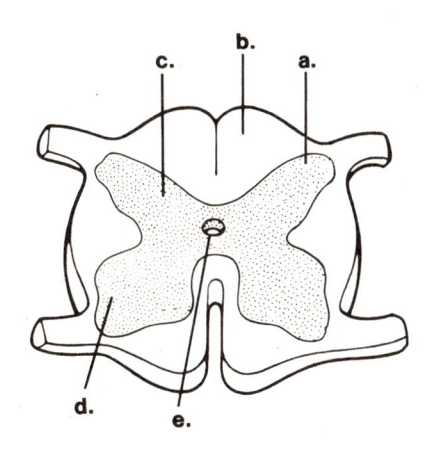

19. Indicate whether the following statements are true (T) or false (F). Change all false statement to true statements.
    _____ a. White matter is white because it contains cell bodies of interneurons that run together in bundles called tracts. Rewrite: _____
    _____ b. The spinal cord carries out the integration of incoming information before sending signals to other parts of the nervous system. Rewrite: _____
    _____ c. When the spinal cord is severed, we suffer a loss of sensation but not a loss of voluntary control. Rewrite: _____

101

20. Label the following drawing of the brain, using the alphabetized list of terms.
    cerebellum, cerebrum, corpus callosum, medulla oblongata, pituitary gland, pons, thalamus

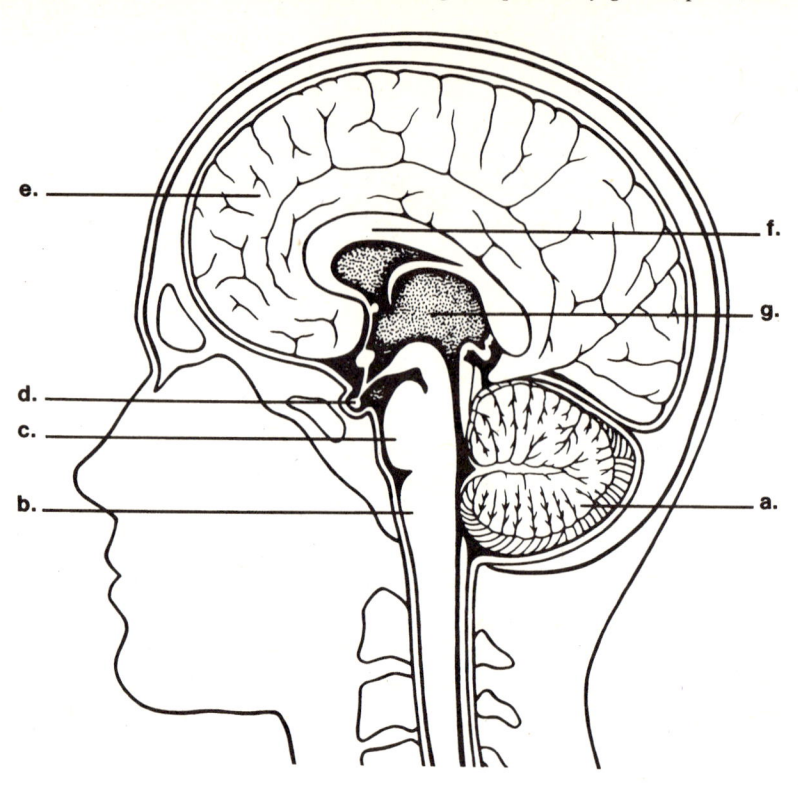

21. Fill in the following table to indicate the functions of the parts of the brain.

Brain Part	Function
Cerebrum	a.
Thalamus	b.
Hypothalamus	c.
Cerebellum	d.
Medulla oblongata	e.

12.4 THE CEREBRAL HEMISPHERE (PAGES 260–261)

- The cerebral cortex contains motor areas, sensory areas, and association areas that are in communication with each other.
- The limbic system contains cortical and subcortical areas that are involved in higher mental functions and emotional responses.

22. Using this alphabetized list, fill in the correct name of the lobe beside the description of functions: *all lobes, frontal lobe, occipital lobe, parietal lobe, temporal lobe.* (Terms can be used more than once.)

    _____ a. Contains primary motor area, which controls voluntary motions.

    _____ b. Contains primary somatosensory area, which receives sensory information from the skin and skeletal muscles.

    _____ c. Contains a primary visual area.

_____ d. Contains a primary auditory area.

_____ e. Contains a primary association area.

_____ f. Carries on higher mental functions like reasoning and critical thinking.

23. Place an X beside those structures that are a part of the limbic system.
    ___ a. tracts that join portions of the cerebral lobes, subcortical nuclei, and the diencephalon
    ___ b. hippocampus, which functions in retrieving memories
    ___ c. amygdala, which adds emotional overtones to memories

24. Which of these best describes the limbic system?
    a. a system that involves reasoning
    b. a system that involves emotions
    c. a system that involves memories
    d. all of these

12.5 HIGHER MENTAL FUNCTIONS (PAGES 262–265)

- Long-term memory depends upon association areas that are in contact with the limbic system.
- There are particular areas in the left hemisphere that are involved in language and speech.

25. Study the following diagram and answer the questions.

    The hippocampus and amygdala are in contact with what areas of the brain: a. _____,
    b. _____, c. _____

    In which of these are memories stored for later retrieval? d. _____

    In which of these areas are memories used to plan future actions? e. _____

    What is the role of the hippocampus? f. _____

    What is the role of the amygdala? g. _____

    Why are there two sets of arrows in the diagram? h. _____

26. Which of these are true statements with regard to language and speech?
    ___ a. Broca's area is a motor speech area.
    ___ b. Damage to Wernicke's area results in the inability to comprehend speech.
    ___ c. Only the left side of the brain contains a Broca's area and a Wernicke's area.

12.6 Homeostasis (page 265)

- Discuss how the nervous system works with other systems of the body to maintain homeostasis.

27. The *Working Together* diagram in your textbook shows how the nervous system aids the other organ systems of the body. Match the organ systems with the correct descriptions.

    a. Brain controls nerves that regulate the heart and dilation of blood vessels.
    b. Receptors send sensory input from bones and joints to brain.
    c. Hypothalamus is part of endocrine system; nerves innervate glands of secretion.
    d. Respiratory centers in brain regulate breathing rate.
    e. Brain controls nerves that innervate smooth muscle and permit digestive tract movements.
    f. Brain controls nerves that regulate size of cutaneous blood vessels, activate sweat glands and arrector pili muscles.
    g. Brain controls nerves that innervate muscles, permitting urination.
    h. Microglial cells engulf and destroy pathogens.
    i. Brain controls onset of puberty; nerves are involved in erection of penis and clitoris, movement of gametes along ducts, and contraction of uterus.
    j. Brain controls nerves that innervate muscles; receptors send sensory input from muscles to brain.

    _____ integumentary system
    _____ lymphatic system
    _____ circulatory system
    _____ skeletal system
    _____ urinary system
    _____ muscular system
    _____ respiratory system
    _____ reproductive system
    _____ endocrine system
    _____ digestive system

12.7 Drug Abuse (pages 265–267)

- Describe drug action in general, and discuss the effects of alcohol, marijuana, cocaine, and heroin.

28. Drugs are believed to affect, in particular, what part of the brain? a. _____ There are both inhibitory and excitatory neurotransmitters in the brain. If a drug blocks the action of an inhibitory neurotransmitter, what psychological effect will it have? b. _____ If a drug blocks the action of an excitatory neurotransmitter, what psychological effect will it have? c. _____

29. The drug a. _____ in tobacco products causes neurons to release dopamine, which reinforces dependence on this drug. Occasional b. _____ users experience euphoria, alterations in vision and judgment, and distortions of space and time. Heavy use of c. _____ often leads to liver damage. d. _____ is a ready-to-smoke form of cocaine.

30. According to endorphin research, what causes heroin withdrawal symptoms? _____
_____

104

Simon Says About Nervous Conduction

For each correct answer, Simon says, "You may move one step forward." Total possible number of steps forward is 10 steps.

1. Which of these would NOT be used when studying nerve conduction?
    a. voltmeter
    b. oscilloscope
    c. electron microscope
    d. electrodes
    e. electric current
2. Which one is NOT *directly* needed for nerve conduction?
    a. dendrites
    b. axons
    c. plasma membrane
    d. nucleus
    e. cytoplasm of the axon
    f. ions
3. Which one does NOT move during nerve conduction?
    a. sodium
    b. potassium
    c. plus charges
    d. minus charges
4. Which one does NOT accurately describe a resting neuron?
    a. positive on the outside of the membrane and negative on the inside
    b. Na$^+$ on the outside of the membrane and K$^+$ on the inside
    c. –65 mV inside
    d. negative on both sides of the membrane
5. Which one is NOT involved with an action potential?
    a. resting potential
    b. permeability
    c. sodium-potassium pump
    d. plasma membrane
    e. acetylcholine
    f. ions
    g. glycogen
6. Which one does NOT conduct a nerve impulse?
    a. sensory neurons
    b. osteocytes
    c. motor neurons
    d. sensory nerves
    e. motor nerves
7. Which one is improperly matched?

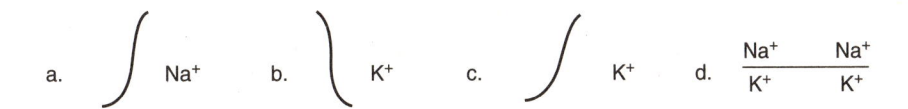

    a. ∫ Na$^+$ b. ⎰ K$^+$ c. ∫ K$^+$ d. $\frac{Na^+ \quad Na^+}{K^+ \quad K^+}$

8. Which number could NOT be associated with an action potential?
    a. –65 millivolts
    b. 0 millivolts
    c. +40 millivolts
    d. –40 watts

9. Which one is improperly matched?
   a. (e⁻) (nerve impulse)
   b. (sodium-potassium pump) (resting potential)
   c. (+) (Na⁺)
   d. (−) (K⁺)
   e. (plasma membrane) (semipermeable)
10. Which one is NOT true?

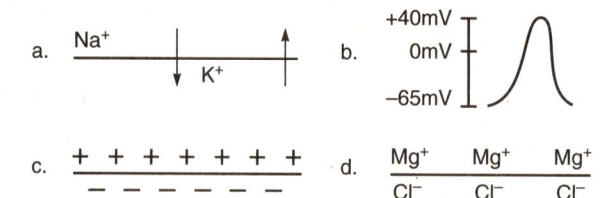

   a. Na⁺ ↓ / K⁺ ↑
   b. +40mV / 0mV / −65mV
   c. + + + + + + / − − − − −
   d. Mg⁺ Mg⁺ Mg⁺ / Cl⁻ Cl⁻ Cl⁻

How many steps were you allowed by Simon? _____

DEFINITIONS WORDSEARCH

Review key terms by completing this wordsearch, using the following alphabetized list of terms:

```
G A N G L I O N P O L I C
C D E T H A L A M U S E C
N G N L L I M B I C Y Y I
S E O K I M D F C H N H T
C E R E B E L L U M A T A
H T U X E L F E R L P N M
U D E N D R I T E O S K O
N K N G O D L O V E E I S
A C E T Y L C H O L I N E
```

acetylcholine
cerebellum
dendrite
ganglion
limbic
neuron
reflex
somatic
synapse
thalamus

a. _____ Portion of brain in third ventricle; integrates sensory input.
b. _____ Nerve cell.
c. _____ Posterior section of brain that coordinates graceful skeletal muscle movement.
d. _____ Fiber of a neuron that conducts signals toward the cell body.
e. _____ Automatic, involuntary response to a stimulus.
f. _____ Region between two neurons where impulses pass.
g. _____ System in brain concerned with emotions and memory.
h. _____ A neurotransmitter.
i. _____ Portion of peripheral nervous system leading to skeletal muscle.
j. _____ Collection of cell bodies within the peripheral nervous system.

CHAPTER TEST

OBJECTIVE TEST

Do not refer to the text when taking this test.

____ 1. Sensory neurons
   a. take impulses to the CNS.
   b. take impulses away from the CNS.
   c. have a cell body in the dorsal root ganglion.
   d. Both *a* and *c* are correct.

____ 2. Which of the following neurons would be found in the autonomic division of the peripheral nervous system?
   a. motor neurons ending in skeletal muscle
   b. motor neurons surrounding the esophagus
   c. sensory neurons at the surface of the skin
   d. sensory neurons attached to olfactory receptors
   e. interneurons in the spinal cord

____ 3. The neuron that is found wholly and completely within the CNS is the
   a. motor neuron.
   b. sensory neuron.
   c. interneuron.
   d. All of these are correct.
____ 4. Which of these contains the nucleus?
   a. axon
   b. dendrite
   c. cell body
   d. Any of these may contain the nucleus.
____ 5. The downswing of the nervous impulse is caused by the movement of
   a. sodium ions to the inside of a neuron.
   b. sodium ions to the outside of a neuron.
   c. potassium ions to the inside of a neuron.
   d. potassium ions to the outside of a neuron.
____ 6. The resting potential is maintained by the sodium-potassium pump.
   a. true
   b. false
____ 7. Rapid conduction of a nerve impulse in vertebrates is due to
   a. the large diameters of the axons.
   b. openings in the myelin sheath.
   c. an abundance of synapses.
   d. the high permeability of neuronal membranes to ions.
   e. All of these are correct.
____ 8. What is involved in a nerve impulse?
   a. ions
   b. electrons
   c. atoms
   d. molecules
____ 9. Synaptic vesicles are
   a. at the ends of dendrites and axons.
   b. at the ends of axons only.
   c. along the lengths of long fibers.
   d. All of these are correct.
____ 10. Acetylcholine
   a. is a neurotransmitter.
   b. crosses the synaptic cleft.
   c. is broken down by acetylcholinesterase.
   d. All of these are correct.
____ 11. A spinal nerve is a
   a. motor nerve.
   b. sensory nerve.
   c. mixed nerve.
   d. All of these are correct.
____ 12. Automatic responses to specific external stimuli require
   a. rapid impulse transmission along the spinal cord.
   b. the involvement of the brain.
   c. simplified pathways called reflex arcs.
   d. the involvement of the autonomic nervous system.
____ 13. Which portion of the nervous system is required for a reflex arc?
   a. mixed spinal nerve
   b. gray matter of spinal cord
   c. cerebrum
   d. Both *a* and *b* are correct.
   e. *a*, *b*, and *c* are correct.
____ 14. The autonomic system has two divisions called the
   a. CNS and peripheral nervous system.
   b. somatic and skeletal systems.
   c. efferent and afferent.
   d. sympathetic and parasympathetic.
____ 15. Motor axons of the somatic system release
   a. acetylcholine.
   b. noradrenalin.
   c. dopamine.
   d. serotonin.
____ 16. Which system is active during stress?
   a. parasympathetic
   b. sympathetic
   c. somatic
   d. All of these are correct.
____ 17. The neurotransmitter of the parasympathetic system is
   a. noradrenalin.
   b. acetylcholine.
   c. cholinesterase.
   d. Both *a* and *b* are correct.
____ 18. Which is the largest part of the human brain?
   a. cerebrum
   b. cerebellum
   c. medulla
   d. thalamus
____ 19. The function of the cerebellum is
   a. consciousness.
   b. motor coordination.
   c. homeostasis.
   d. sense reception.
____ 20. Which portion of the brain is involved in judgment?
   a. cerebellum
   b. frontal lobe of cerebrum
   c. medulla
   d. parietal lobe of cerebrum
____ 21. The drug that is classified as a hallucinogen is
   a. marijuana.
   b. alcohol.
   c. caffeine.
   d. nicotine.
____ 22. Drugs of abuse primarily affect the
   a. cerebellum.
   b. medulla oblongata.
   c. limbic system.
   d. thalamus.
____ 23. A ready-to-smoke, highly addictive form of cocaine is
   a. heroin.
   b. crack.
   c. marijuana.
   d. alcohol.

THOUGHT QUESTIONS

Answer in complete sentences.

24. Contrast the way the nerve impulse travels along an axon with the way it travels across a synapse.

25. In either of what two ways would you expect an inhibitory psychoactive drug to affect transmission across a synapse?

Test Results: _____ number correct ÷ 25 = _____ × 100 = _____ %

ANSWER KEY

STUDY QUESTIONS

1. a. sends signal to cell body **b.** control center **c.** takes impulse away from cell body **2. a.** sensory receptor **b.** cell body **c.** axon **d.** dendrites **e.** axon **f.** dendrite **g.** axon **h.** nucleus of Schwann cell **i.** node of Ranvier **j.** effector **3.** to take nerve impulses to CNS **4.** to take nerve impulses from one part of CNS to another **5.** to take nerve impulses away from CNS **6. a.** plus on outside; minus on inside **b.** Na^+ on outside and K^+ on inside **c.** minus on outside; plus on inside **d.** Some Na^+ have moved to inside **e.** plus on outside; minus on inside **f.** Some K^+ have moved to outside. **7. a.** +40 **b.** Na^+ (sodium ions) gates open **c.** –65 **d.** K^+ (potassium ions) gates open **e.** –65 **8.** sodium-potassium **9. a.** synapse **b.** synaptic vesicles **c.** axon **d.** dendrite **e.** postsynaptic membrane **f.** synaptic cleft **g.** neurotransmitter **10.** reception of neurotransmitter at receptor site **11.** b, d **12. a.** cranial **b.** spinal **c.** bundle of fibers (axons) **d.** it contains both sensory and moter fibers **13. a.** sensory fibers **b.** sensory fiber **c.** dorsal root gaglion **d.** interneuron **e.** motor fiber **f.** effector **14. a.** receptor **b.** sensory **c.** internal organs **b.** have motor neurons **c.** have ganglia **17. a.** fight or flight **b.** normal activity **c.** norepinephrine (NE) **d.** acetylcholine (ACh) **e.** near cord **f.** near organ **g.** spinal nerves only **h.** spinal nerves plus vagus **18. a.** dorsal horn **b.** white matter **c.** gray matter **d.** ventral horn **e.** central canal **19. a.** F...contains myelinated axons... **b.** T **c.** F...and a loss of voluntary control... **20. a.** cerebellum **b.** medulla oblongata **c.** pons **d.** pituitary gland **e.** cerebrum **f.** corpus callosum **g.** thalamus **21. a.** motor control, higher levels of thought **b.** integrates and sends sensory information to cerebrum **c.** homeostasis **d.** motor coordination **e.** control of internal organs **22. a.** frontal lobe **b.** parietal lobe **c.** occipital lobe **d.** temporal lobe **e.** all lobes **f.** frontal lobe **23.** a, b, c **24.** d **25. a.** cortical sensory areas **b.** prefrontal area **c.** diencephalon **d.** cortical sensory areas **e.** prefrontal area **f.** serves as bridge between prefrontal area and cortical sensory areas **g.** adds emotional overtones to memories **h.** one set of arrows for semantic memory and one set for episodic memory **26.** a, b, c **27.** f, h, a, b, g, j, d, i, c, e **28. a.** limbic system **b.** increased likelihood of excitation **c.** decrease likelihood of excitation **29. a.** nicotine **b.** marijuana **c.** alcohol **d.** Crack **30.** Body's production of endorphins has decreased.

SIMON SAYS ABOUT NERVOUS CONDUCTION

1. c **2.** d **3.** d **4.** d **5.** g **6.** b **7.** c **8.** d **9.** a **10.** d

DEFINITIONS WORDSEARCH

```
    G A N G L I O N
        T H A L A M U S   C
      N   L I M B I C Y   I
      O           N       T
    C E R E B E L L U M A   A
      U X E L F E R   P   M
      D E N D R I T E S   O
      N               E   S
    A C E T Y L C H O L I N E
```

a. thalamus **b.** neuron **c.** cerebellum **d.** dendrite **e.** reflex **f.** synapse **g.** limbic **h.** acetylcholine **i.** somatic **j.** ganglion

CHAPTER TEST

1. e **2.** b **3.** c **4.** c **5.** d **6.** a **7.** b **8.** a **9.** b **10.** d **11.** c **12.** c **13.** d **14.** d **15.** a **16.** b **17.** b **18.** a **19.** b **20.** b **21.** a **22.** c **23.** b **24.** There is an exchange of Na^+ and K^+ as the nerve impulse travels along an axon, but the release of a neurotransmitter causes the nerve impulse to travel across a synapse. **25.** An inhibitory psychoactive drug could either prevent the action of an excitatory neurotransmitter or promote the action of an inhibitory neurotransmitter at a synapse.

13
SENSES

STUDY TIPS

Sensory receptors are specialized structures designed to convey information to the central nervous system about the environment. They help our bodies to maintain homeostasis.

In this chapter, you will first learn about the interesting variety of receptors in the human body (p. 272). The skin has receptors that respond to five different types of stimuli (p. 275). The special senses, associated with the head, are the most complex and fascinating of the senses. The senses of taste and smell both employ chemoreceptors (p. 276) to detect the quality of food and to determine other characteristics of the environment. Eyes are intriguing organs, designed to perceive light (p. 278–279). Vision is possible because of visual pigments (p. 280) of photoreceptors that become altered when light strikes them, generating a nervous impulse.

When you study the eye, make a diagram like Figure 13.7, and label all the parts of the eye. Next, make note of the functions of each portion of the eye (p. 278). Pay close attention to how the retina is structured (p. 279) because that will help you understand why photoreceptors work the way they do (p. 280).

The ear houses the senses of hearing and equilibrium, making use of mechanoreceptors in both cases. Sketch the anatomy of the ear (see Figure 13.13), labeling the parts and their functions. Be sure to include the anatomy and functioning of the inner ear (see Fig. 13.14), where hair cells for hearing and equilibrium are located.

STUDY QUESTIONS

Study the text section by section as you answer the questions that follow.

13.1 SENSORY RECEPTORS AND SENSATIONS (PAGES 272–273)

- Describe the various receptors, and state the type of stimuli they receive.

1. Match the following statements to the terms listed in the key.
    Key chemoreceptors mechanoreceptors proprioceptors thermoreceptors pain receptors photoreceptors

    _____ a. located only in the eye
    _____ b. monitor the pH of the blood
    _____ c. detect tissue damage
    _____ d. detect stretch in tendons and ligaments
    _____ e. sensitive to changes in heat and cold
    _____ f. hearing
    _____ g. taste and smell

2. Name two types of proprioceptors: a. _____ and b. _____
   When a muscle spindle relaxes, c. _____
   Muscle spindles have two functions, d. _____ and e. _____

109

13.2 SOMATIC SENSES (PAGES 274–275)

- Discuss the general receptors of the skin.

Indicate whether the following statements are true (T) or false (F). Rewrite the false statements in questions 3–10 to make true statements.

3. _____ There are sensory receptors in the epidermis, dermis, and subcutaneous layers. Rewrite:_____

4. _____ Each sensory receptor is stimulated by various stimuli, such as heat, cold, touch, pressure, and pain. Rewrite: _____

5. _____ When stimulated, a sensory receptor generates a nerve impulse that travels in a sensory neuron to the CNS. Rewrite: _____

6. _____ Each type of sensory receptor is sensitive to only one stimulus. Rewrite: _____

7. _____ Sensation occurs in the brain and not at the sensory receptor. Rewrite: _____

8. _____ Sensory receptors generate nerve impulses. Rewrite: _____

9. _____ Sensory receptors are uniformly distributed throughout the skin. Rewrite: _____

10. _____ Sensory receptors are part of a reflex arc. Rewrite: _____

11. Match the types of receptors with the sense they detect.
    a. pressure   b. heat   c. cold   d. pain   e. touch

    _____ free nerve endings
    _____ Merkel disks
    _____ Krause end bulbs
    _____ Meissner corpuscles
    _____ Pacini corpuscles
    _____ Ruffini endings

13.3 CHEMICAL SENSES (PAGES 276–277)

- Describe the senses that rely on chemoreceptors.
- Explain how chemoreceptors operate.

12. Match items in the key to the statements that follow.
    Key taste receptors   smell receptors   both taste and smell receptors

    _____ a. receptor proteins combine with chemical
    _____ b. brain senses impulses as a weighted average
    _____ c. taste buds with microvilli house receptor proteins
    _____ d. salty receptor proteins on tip of tongue
    _____ e. olfactory cells
    _____ f. are not effective when you have a cold
    _____ g. easily adapt to outside stimuli
    _____ h. involved in enjoyment of food
    _____ i. a characteristic combination of receptor proteins are activated

13. The senses of taste and smell work because specific a._____ in the organs of taste and smell combine with b._____ in the air or food. Both senses thus employ c._____ to detect changes in the environment.

13.4 SENSE OF VISION (PAGES 278–283)

- Describe the anatomy of the eye and the function of each part.
- Describe the receptors for sight, their mechanism of action, and the mechanism for stereoscopic vision.
- Identify common disorders of sight discussed in the text.

14. Using the alphabetized list of terms and the answer blanks provided, state the name and function of each part of the eye indicated in the illustration.

    choroid
    ciliary body
    cornea
    fovea centralis
    iris
    lens
    optic nerve
    retina
    sclera

    Structure   **Function**

    a. _____   _____
    b. _____   _____
    c. _____   _____
    d. _____   _____
    e. _____   _____
    f. _____   _____
    g. _____   _____
    h. _____   _____
    i. _____   _____

15. The lens is a._____ for distant objects and b._____ for close objects. This is called c._____.

16. The receptors of sight are classified as a._____. Two kinds exist: b._____ perceive motion and are responsible for night vision, and c._____ are responsible for color vision.

17. Rods have a pigment called a._____, which is made up of the protein b._____ and a pigment molecule called c._____, a derivative of vitamin A. When light strikes the pigment molecule, the rhodopsin is activated.

    Color vision depends on d._____ kinds of cones, each of which has a slightly different structure of e._____ molecule. Each is able to detect a different wavelength, or color, of light.

    Rods are located throughout the retina, but cones are concentrated in the f._____.

18. Considering the layers of the retina—rods and cones/bipolar cells/ganglion cells—which of these are at the back of the retina (closest to the choroid)? a._____ Which of these are receptors for sight? b._____ Which of these are fewest in number? c._____

    This supports the belief that d._____ occurs in the retina.

19. With reference to the following figure, the region where the optic nerves cross is the a._____. Each primary visual area of the cerebral cortex receives information about (choose *the complete* or *one-half the*) b._____ visual field. Also, it is now known that the visual areas (choose *parcel out* or *retain as a unified whole*) c._____ information regarding color, form, motion, etc. This means that the cerebral cortex has to (choose *rebuild* or *imagine*) d._____ the visual field before we can "see" it.

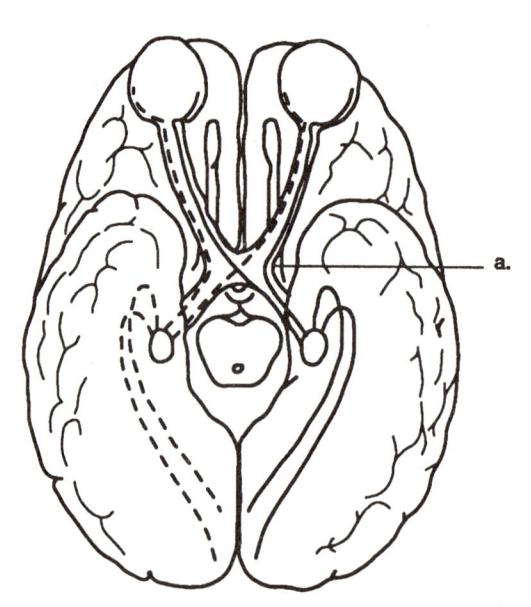

20. Fill in the blanks in this table.

Name	Description	Image Focused	Correction
Nearsightedness	See nearby objects	a.	Concave lens
Farsightedness	b.	c.	d.
Astigmatism	Cannot focus	Image not focused	e.

13.5 SENSE OF HEARING (PAGES 286—288)

- Describe the anatomy of the ear and the function of each part.
- Discuss the receptors for balance and hearing and their mechanism of action.
- Identify the two types of deafness.

21. Using the alphabetized list of terms and the answer blanks provided, state the name and function of each part of the ear indicated in the illustration.

    auditory canal
    auditory tube
    cochlea
    cochlear nerve
    (malleus) hammer
    pinna
    semicircular canal
    (stapes) stirrup
    tympanic membrane
    vestibule

    Structure/Function

    a. _____
    c. _____
    e. _____
    g. _____
    i. _____

    Structure/Function

    b. _____
    d. _____
    f. _____
    h. _____
    j. _____

22. Place each of the following statements in proper sequence by using the numbers 1-9 to describe the events that lead to the formation of an auditory nerve impulse.
    _____ a. vibration is transferred from the hammer to the anvil to the stirrup
    _____ b. basilar membrane moves up and down
    _____ c. nerve impulse is transmitted in cochlear nerve to brain
    _____ d. sound waves pass through the auditory canal
    _____ e. cilia of hair cells rub against tectorial membrane
    _____ f. sound waves cause tympanic membrane to vibrate
    _____ g. nerve impulse is generated
    _____ h. vibrations move from vestibular canal to tympanic canal
    _____ i. membrane at oval window vibrates

13.6 SENSE OF BALANCE (PAGE 289)

- The sensory receptors for balance are hair cells in the vestibule and semicircular canals of the inner ear, which respond to the tilt of the head and the movement of the body, respectively.

23. For a. _____ equilibrium, angular and/or rotational movement occurs such as when the body is moving. Fluid in the semicircular canals displaces the gelatinous material within the b. _____. For c. _____ equilibrium, movement occurs in one plane (vertical or horizontal), and the otoliths in the d. _____ and saccule are displaced in the gelatinous material, bending the cilia of the hair cells. Hair cells in the ampullae and utricle and saccule synapse with the e. _____ nerve.

DEFINITIONS WORDSEARCH

Review key terms by completing this wordsearch, using the following alphabetized list of terms:

```
A C C O M M O D A T I O N X
M R E H T O P S D N I L B V
P E T U O R E B Y U J D F E
U J I K S R R E D S I C L I
L H U T S C O C H L E A L O
L G Y C I S X I W A S N I U
A B H I C L O P D S I D E D
K O T O L I T H B C Z T Y U
M U J K E R H O D O P S I N
```

accommodation
ampulla
blind spot
choroid
cochlea
ossicle
otolith
rhodopsin

a. _____ Lens adjustment to see close objects.
b. _____ Vascular, pigmented layer of eyeball.
c. _____ Snail-shell portion of inner ear.
d. _____ Area of eye where optic nerve passes through retina.
e. _____ Base of semicircular canal in inner ear.
f. _____ One of the small bones of the middle ear.
g. _____ Calcium carbonate granule within inner ear.
h. _____ Visual pigment found in rods.

CHAPTER TEST

OBJECTIVE TEST

Do not refer to the text when taking this test.

____ 1. A receptor
   a. is the first portion of a reflex arc.
   b. initiates a nerve impulse.
   c. responds to only one type of stimulus.
   d. is attached to a dendrite.
   e. All of the above are correct.

____ 2. There are five sense receptors in the skin. They are
   a. choroid, cochlea, ossicles, and otoliths.
   b. hot, cold, pressure, touch, and pain.
   c. sclera, choroid, cornea, cones, and cochlea.
   d. mechanoreceptors, chemoreceptors, and photoreceptors.

____ 3. Taste cells and olfactory cells are both
   a. somatic senses.
   b. mechanoreceptors.
   c. pseudociliated epithelium.
   d. chemoreceptors.

____ 4. The blind spot is
   a. a nontransparent area on the lens.
   b. a nontransparent area on the cornea.
   c. on the retina, where there are no rods or cones.
   d. called the fovea centralis.

Match the items in the key to questions 5–8.
   Key a. retina   b. optic nerve   c. lens   d. all   e. none

____ 5. Which of these is (are) necessary to proper vision?
____ 6. Which of these contain(s) receptors for sight?
____ 7. Which of these focus(es) light?
____ 8. In which is the sensation of sight realized?
____ 9. The current theory of color vision proposes that
   a. there are three primary colors associated with color vision.
   b. cones respond selectively to different wavelengths of light.
   c. the rods are responsible for nighttime color vision.
   d. Both *a* and *b* are correct.
   e. *a*, *b*, and *c* are correct.

____ 10. Both sides of the brain receive information from both eyes due to the
   a. optic chiasma.
   b. fovea centralis.
   c. ciliary muscle.
   d. ciliary body.

____11. If you are nearsighted, the image is focused
   a. in front of the retina.
   b. behind the retina.
   c. on the retina.
   d. at the blind spot.
____12. The disorders of nearsightedness and farsightedness are due to
   a. an eyeball of incorrect length.
   b. a cloudy lens.
   c. pressure increase.
   d. a torn retina.
____13. Vitamin A is needed for the
   a. lens.
   b. rods.
   c. cones.
   d. cornea.
____14. The cochlear nerve is associated with the
   a. spiral organ.
   b. ossicles.
   c. tympanic membrane.
   d. auditory tube.

Match the items in the key to questions 15–17.
   Key a. ossicles   b. otoliths   c. cochlea
           d. auditory tube

____15. Which of these have nothing to do with hearing?
____16. In which would you find receptors for hearing?
____17. Which of these is concerned with balance?

____18. Balance receptors differ from hearing receptors in that balance receptors _____, while hearing receptors _____.
   a. are located in the outer ear; are located in the inner ear
   b. respond to pressure waves; do not respond to pressure waves
   c. consist of hair cells; do not consist of hair cells
   d. are in the semicircular canals; are in the spiral organ
____19. Which part of the ear is for balance?
   a. outer
   b. middle
   c. inner
   d. All of these are correct.
____20. Nerve deafness may be due to
   a. fused ossicles.
   b. worn stereocilia.
   c. German measles.
   d. Both *a* and *c* are correct.

Match the items in the key to questions 21–25.
   Key a. organ of Corti   b. rods and cones
           c. pressure receptor   d. olfactory receptor
           e. taste buds   f. utricle and saccule

____21. Which of these are *not* mechanoreceptors?
____22. Which of these is located in the skin?
____23. Which of these help maintain equilibrium?
____24. Which of these are located in the ear?
____25. Which of these are stimulated by chemicals?

Thought Questions

Answer in complete sentences.
26. The length of the organ of Corti is related to what ability in humans?

27. Relate the function of the rods to their placement in the retina.

Test Results: _____ number correct ÷ 27 = _____ × 100 = _____%

Answer Key

Study Questions

1. a. photoreceptors **b.** chemoreceptors **c.** pain receptors **d.** proprioceptors, mechanoreceptors **e.** thermoreceptors **f.** mechanoreceptors **g.** chemoreceptors **2. a.** muscle spindle **b.** Golgi tendon organ **c.** nerve impulses are generated **d.** maintain posture **e.** tell position of limbs **3.** F. There are receptors only in the dermis. **4.** F. Each receptor is stimulated by only one type of stimulus. **5.** T **6.** T **7.** T **8.** T **9.** F. Some areas of skin have more of a particular type of receptor and less of others. **10.** T **11.** d, e, c, e, a, b **12. a.** both **b.** taste receptors **c.** taste receptors **d.** taste receptors **e.** smell receptors **f.** smell receptors **g.** smell receptors **h.** both **i.** small receptors **13. a.** protein receptors **b.** chemicals (molecules) **c.** chemoreceptors **14. a.** retina; photoreceptors for sight **b.** fovea centralis; makes acute vision possible **c.** ciliary body; holds lens in place; accommodation **d.** sclera; protects eyeball **e.** choroid; absorbs stray light rays **f.** optic nerve; transmission of nerve impulse **g.** lens; focusing **h.** cornea; refracts light rays **i.** iris; regulates entrance of light **15. a.** flat **b.** rounded **c.** accommodation **16. a.** photoreceptors **b.** rods **c.** cones **17. a.** rhodopsin **b.** opsin **c.** retinal **d.** three **e.** opsin **f.** fovea centralis **18. a.** rods and cones **b.** rods and cones **c.** ganglion cells **d.** integration **19. a.** optic chiasma **b.** one-half the **c.** parcel out **d.** rebuild **20. a.** in front of retina **b.** see distant objects **c.** behind retina **d.** convex lens **e.** irregular lens **21. a.** tympanic membrane; starts vibration of ossicles **b.** semicircular canal; dynamic equilibrium **c.** cochlea; contains hearing receptors **d.** cochlear nerve; transmission of nerve impulse **e.** (malleus) hammer; transmits vibrations **f.** pinna; reception of sound waves **g.** auditory canal; collection of sound waves **h.** (stapes) stirrup; transmits vibrations to oval window **i.** auditory tube; connects middle ear to pharynx **j.** vestibule; static equilibrium **22.** d, f, a, i, h, b, e, g, c **23. a.** dynamic **b.** cupula **c.** static **d.** utricle **e.** vestibular

Definitions Wordsearch

```
A C C O M M O D A T I O N
M   H T O P S D N I L B
P       O
U       S R
L       S C O C H L E A
L       I   I
A       C   D
  O T O L I T H
    E R H O D O P S I N
```

a. accommodation **b.** choroid **c.** cochlea **d.** blind spot **e.** ampulla **f.** ossicle **g.** otolith **h.** rhodopsin

Chapter Test

1. e **2.** b **3.** d **4.** c **5.** d **6.** a **7.** c **8.** e **9.** d **10.** a **11.** a **12.** a **13.** b **14.** a **15.** b and d **16.** c **17.** b **18.** d **19.** c **20.** b **21.** b, d, e **22.** c **23.** f **24.** a, f **25.** d, e **26.** The length of the organ of Corti is related to the ability of humans to hear a range of pitches. **27.** The rods are located throughout the retina, and this is consistent with their ability to detect motion.

14
ENDOCRINE SYSTEM

STUDY TIPS

The endocrine system is one of the two principal controlling systems of the body. As you learned in chapter 12, the nervous system is the other. Chemical messengers known as hormones (p. 294) are released from endocrine glands and travel throughout the bloodstream until a target cell detects them.

Hormones work in one of two ways when a target cell detects their presence. Steroid hormones (p. 295) activate DNA and alter target cell activity at the level of the gene. Nonsteroid hormones (p. 295) activate second messengers that turn on a host of enzymes within the cell.

Many endocrine glands are distributed throughout the body. Construct a detailed chart (see Table 14.1 in the text) listing the name of the endocrine gland, its location, the hormone(s) it releases, and the functions of those hormones in the body. Pay particular attention to those hormones that oppose each other (e.g., insulin and glucagon) in controlling levels of bodily substances.

STUDY QUESTIONS

Study the text section by section as you answer the questions that follow.

14.1 ENVIRONMENTAL SIGNALS (PAGES 294–297)

- Discuss the concept of an environmental signal, and give three categories for environmental signals.

1. Give an example of an environmental signal that acts at a distance between individuals.
   a._____

   Give an example of an environmental signal that acts at a distance between body parts.
   b._____

   Give an example of an environmental signal that acts locally between adjacent cells.
   c._____

2. Which is both a neurotransmitter and a hormone?
   _____ a. cortisone
   _____ b. norepinephrine
   _____ c. ACTH
   _____ d. thyroxine

3. Write either *peptide hormone* or *steroid hormone* on the line above each diagram. Using the alphabetized list of terms, place an appropriate word or phrase on the lines within each diagram.
   active, cyclic AMP, hormone receptor, hormone receptor complex, protein synthesis

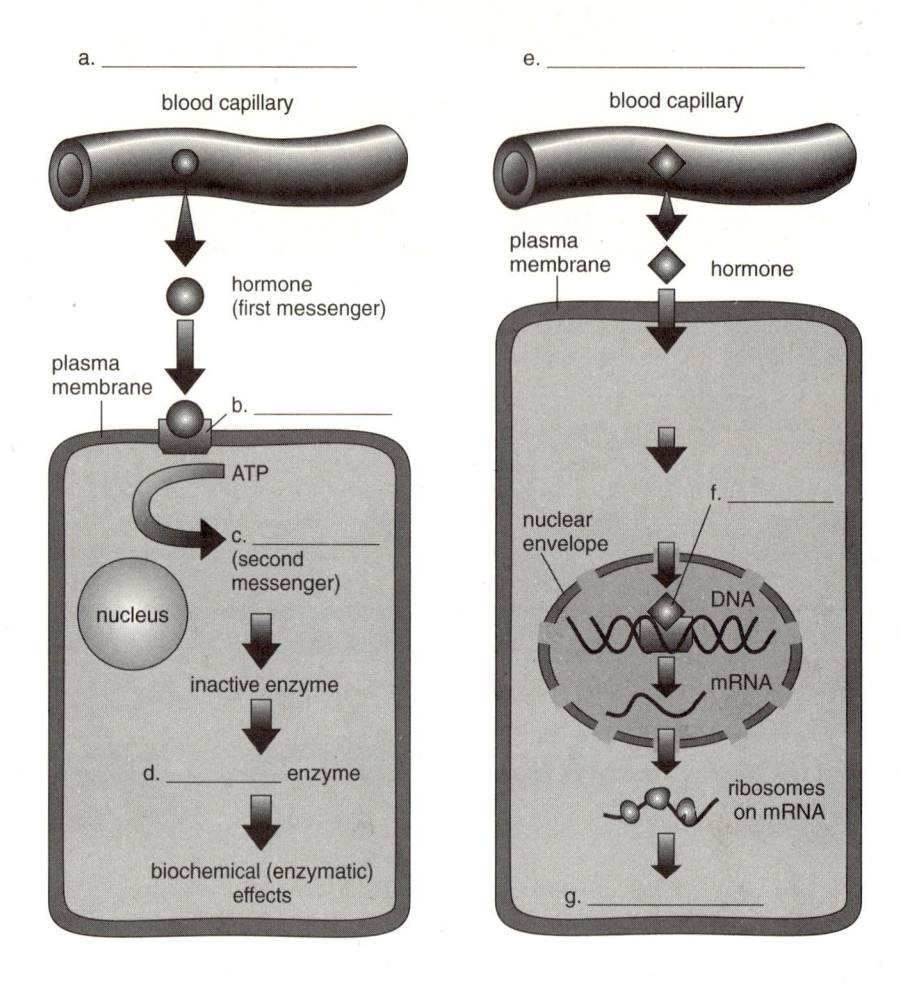

a. __peptide hormone__

b. __hormone receptor__

c. __cyclic AMP__ (second messenger)

d. __active__ enzyme

e. __steroid hormone__

f. __hormone receptor complex__

g. __protein synthesis__

Endocrine Glands (Page 297)

4. Identify the endocrine glands in the following diagram, and name at least one hormone produced from each.

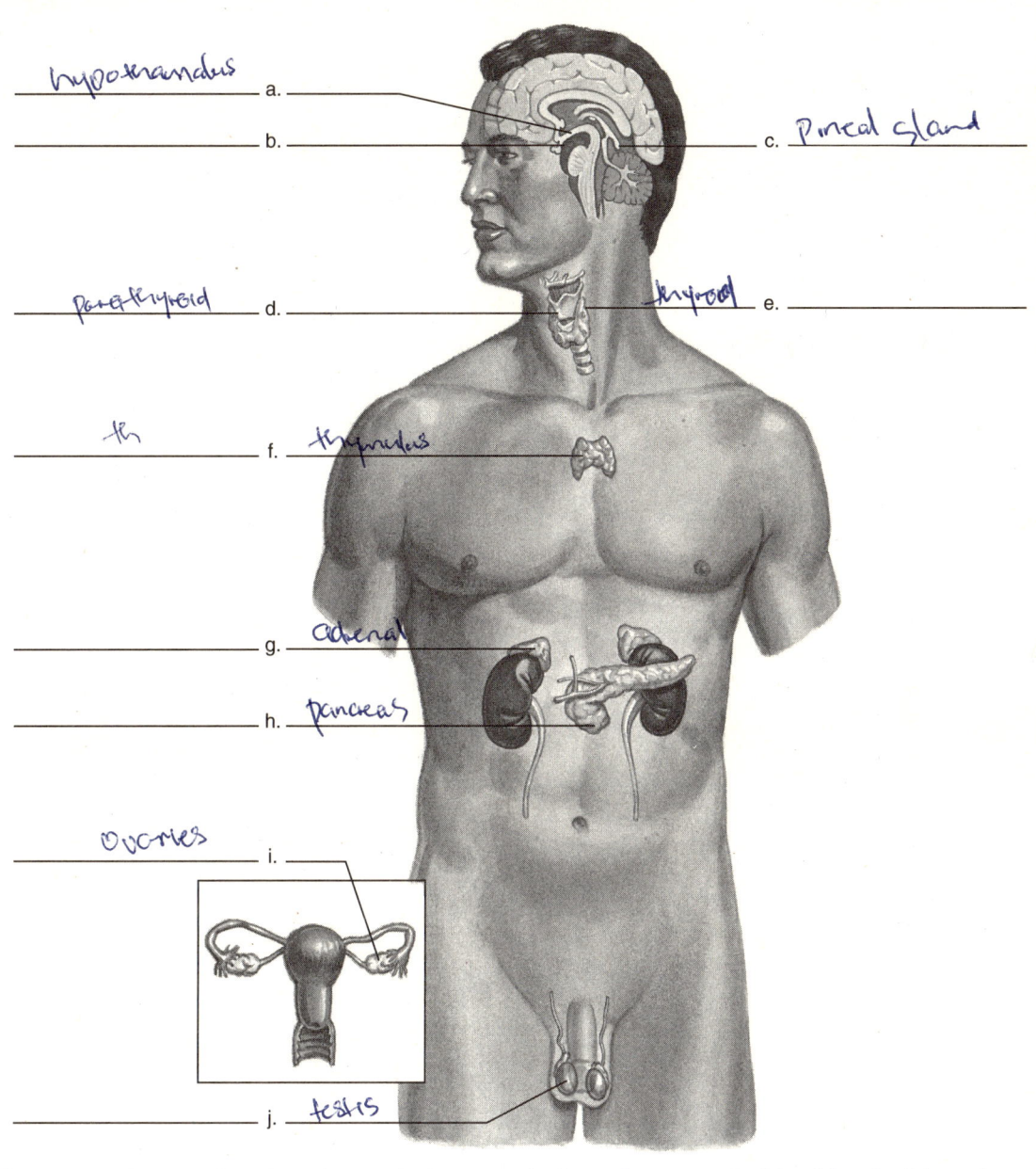

a. hypothalamus
b. _____
c. Pineal gland
d. parathyroid
e. thyroid
f. thymus
g. adrenal
h. pancreas
i. ovaries
j. testis

5. Control of hormone release. Often the release of a hormone is dependent upon the blood level of the substance it is controlling. When the level of a substance increases, generally this causes the hormone secretion to a. _____. This is an example of b. _____ feedback. In other instances, c. _____ hormones oppose each other's actions, thus regulating the target substance in the body.

14.2 Hypothalamus and Pituitary Gland (pages 298–300)

- Contrast the ways in which the hypothalamus controls the posterior and anterior pituitary.
- List the hormones produced by the anterior and posterior pituitary, and describe their effects.

6. Add dotted lines to this line diagram wherever feedback control is usually exerted.

Hypothalamus—Releasing Hormone—Anterior Pituitary Hormone—Gland—Hormone
       (1)   (2)   (3)

7. If the "gland" in the diagram in question 6 is the thyroid, then the first hormone is a._____, the second is b._____, and the third is c._____. If the "gland" in the diagram in question 6 is the adrenal cortex, then the first hormone is d._____, the second is e._____, and the third is f._____.

8. Write *AP* beside the statements that apply to the anterior pituitary and *PP* beside the statements that apply to the posterior pituitary.
   ____ a. connected to hypothalamus by nerve fibers
   ____ b. connected to hypothalamus by blood vessels
   ____ c. secretes hormones produced by hypothalamus
   ____ d. controlled by releasing hormones produced by hypothalamus

9. To show why the anterior pituitary is sometimes called the master gland, complete this table.

Anterior Pituitary Produces	Gland Controlled	Hormone Produced by Gland
TSH	a. Thyroid	b. Thyroxine
ACTH	c. Adrenal	d. adrenaline
Gonadotropic hormones		
Female	e.	f.
Male	g.	h.

10. What causes the hypothalamus to stop producing releasing hormones and the anterior pituitary to stop producing its hormones? _____

11. The anterior pituitary produces three other hormones. The hormone a. _prolactin_ causes the mammary glands to develop and produce milk. b. _____ hormone causes skin color changes in lower vertebrates. Growth hormone (GH) promotes cell division, protein synthesis, and c. _____ growth. If too little GH is produced during childhood, the individual becomes a pituitary d. _____. If too much is produced, the individual is a pituitary e. _____. If there is an overproduction of GH in the adult, f. _____ results, and the face, hands and feet g. _____.

14.3 THYROID AND PARATHYROID GLANDS (PAGES 301–302)

- List the hormones of the thyroid gland, and give their functions.
- Explain why iodine is a necessary dietary component for thyroid hormones.
- Give the function of parathyroid hormone.
- Be able to identify thyroid disorders.

12. Match the conditions in the key with the phrases that follow.
    Key cretinism   exophthalmic goiter   simple goiter   myxedema
    hypothyroidism (choose more than one) a._____
    hyperthyroidism b._____
    hypothyroidism since birth c._____
    hypothyroidism in the adult d._____
    lack of iodine e._____

13. Match the items in the key with the phrases that follow.
    Key Ca++ in bones   Ca++ in blood

    _____ a. calcitonin present
    _____ b. mistaken removal of parathyroids during a thyroid operation
    _____ c. PTH present
    _____ d. postmenopausal women who no longer produce estrogen
    _____ e. osteoporosis
    _____ f. child with tetany

14.4 ADRENAL GLANDS (PAGES 303–305)

- Describe the hormones of the adrenal medulla, and explain how their release is controlled.
- List the hormones of the adrenal cortex, their functions, and how their release is controlled.

14. What hormones are produced by the adrenal medulla, and what effect do they have on the body?
    a. _____

    What triggers the release of hormones from the adrenal medulla? b. _____

15. Distinguish between cortisol and aldosterone by writing *yes* or *no* on each line.

	Cortisol	Aldosterone
controlled by ACTH	a._____	_____
glucocorticoids	b._____	_____
mineralocorticoid	c._____	_____
relieves stress	d._____	_____
Na+/K+ balance	e._____	_____
amino acids → glucose	f._____	_____
controlled by angiotensin II	g._____	_____

16. Indicate whether these symptoms are indicative of Addison disease or Cushing syndrome by writing *AD* or *CS* beside each description.
    _____ a. cannot handle bodily stress
    _____ b. cannot maintain blood glucose level
    _____ c. tendency toward diabetes mellitus
    _____ d. low blood pressure because of a low blood sodium level
    _____ e. high blood pressure because of a high blood sodium level
    _____ f. edema because of too much sodium in system
    _____ g. bronzing of skin
    _____ h. thin arms and legs; enlarged trunk

14.5 PANCREAS (PAGES 306–307)

- Describe the functions of insulin and glucagon.
- Discuss the problems associated with diabetes.

17. Write the word *insulin* or *glucagon* on the appropriate arrow.
    glycogen a. ←___insulin___ glucose molecules
    storage in liver b. ___glucagon___→ in the blood

18. Complete each of the following statements with the term *increases* or *decreases*:
    Glucagon a. ___increases___ blood sugar concentration. In type I diabetes, insulin production from the
    pancreas b. _____. In type II diabetes, the response of body cells to the influence of
    insulin c. _____.

121

14.6 OTHER ENDOCRINE GLANDS (PAGES 309–310)

- List the functions of other endocrine glands.
- Give examples of medical conditions associated with the overproduction or underproduction of hormones discussed in this chapter.

19. The pineal gland produces the hormone a._____ that helps regulate b._____. The ovaries produce c._____ and d._____ that function in the menstrual cycle and during pregnancy. The testes produce e._____ that influences the production of f._____ and secondary sexual characteristics.

20. Match the items in the key to the phrases that follow.
    Key acromegaly osteoporosis diabetes insipidus Cushing disease diabetes mellitus Addison disease exophthalmic goiter

    enlarged thyroid and bulging eyes a._____
    bronzing of skin b._____
    weak bones c._____
    enlarged bones d._____
    sugar in the urine e._____
    thyroid f._____
    adrenal cortex g._____ and _____
    pancreas h._____
    posterior pituitary i._____

14.7 HOMEOSTASIS (PAGE 310)

- Discuss how the endocrine system works with other systems of the body to maintain homeostasis.

21. The *Working Together* diagram in your textbook shows how the endocrine system benefits from the actions of the other organ systems of the body. Match the organ systems with the correct descriptions.
    a. *Lymphatic vessels pick up excess tissue fluid; immune system protects against infections.*
    b. *Bones provide protection for glands; store Ca^{++} used as second messenger.*
    c. *Kidneys keep blood values within normal limits so that transport of hormones continues.*
    d. *Stomach and small intestines produce hormones.*
    e. *Gas exchange in lungs provides oxygen and rids body of carbon dioxide.*
    f. *Hypothalamus is part of endocrine system; nerves innervate glands of secretion.*
    g. *Gonads produce sex hormones.*
    h. *Muscles help protect glands.*
    i. *Skin helps protect endocrine glands.*
    j. *Blood vessels transport hormones from glands; blood services glands; heart produces atrial natriuretic hormones.*

    _____ integumentary system
    _____ lymphatic system
    _____ circulatory system
    _____ skeletal system
    _____ urinary system
    _____ muscular system
    _____ respiratory system
    _____ reproductive system
    _____ endocrine system
    _____ digestive system

Hormone Hockey

For every 5 correct answers in sequence, you have scored one goal.

First goal: Match the hormone to the glands (a–i).

Glands:
- a. anterior pituitary
- b. thyroid
- c. parathyroids
- d. adrenal cortex
- e. adrenal medulla
- f. pancreas
- g. gonads
- h. pineal gland
- i. posterior pituitary

Hormones:
- _f_ 1. insulin
- _i_ 2. oxytocin
- _c_ 3. melatonin
- _d_ 4. cortisol
- _b_ 5. thyroxine

Second goal: Match the condition to the glands (a–i).

Conditions:
- ____ 6. diabetes mellitus
- ____ 7. creatinism
- ____ 8. Addison disease
- ____ 9. hypertension
- ____ 10. giant

Third goal: Match the function to the hormones.
- a. melatonin
- b. estrogens
- c. androgens
- d. insulin
- e. glucagon
- f. epinephrine
- g. aldosterone
- h. cortisol
- i. parathyroid hormone
- j. thyroxine
- k. calcitonin (lowers)
- l. antidiuretic hormone

Functions
- ____ 11. raises blood calcium level
- _j_ 12. reduces stress
- ____ 13. maintains secondary female sex characteristics
- ____ 14. involved in circadian rhythms
- ____ 15. stimulates water reabsorption by kidneys

Fourth goal: Match the glands to the hormones (a–l). Some glands require two answers.

Glands
- ____ 16. testes *estrogen progesteron*
- ____ 17. adrenal cortex
- ____ 18. pancreas *insulin*
- ____ 19. thyroid *thyroxine*
- ____ 20. adrenal medulla *adrenaline*

Fifth goal: Select five hormones secreted by the anterior pituitary by answering yes or no to each of these.

- ____ 21. thyroid-stimulating hormone
- ____ 22. androgens
- ____ 23. gonadotropic hormones
- ____ 24. glucagon
- ____ 25. oxytocin
- ____ 26. growth hormone
- ____ 27. prolactin
- ____ 28. antidiuretic hormone
- ____ 29. estrogens
- ____ 30. adrenocorticotropic hormone

How many goals did you make? _____

123

DEFINITIONS WORDSEARCH

Review key terms by completing this wordsearch, using the following alphabetized list of terms:

```
C R E T I N I S M N U J L I P
O A R H G T Z B E S A I L H R
R A L P O X Y T O C I N H U O
T M L C F R S H U K N H S R S
I E R G I U M N V Y S A F T T
S D G T R T D O F W U J H U A
O E P O L I O U N Y L R E R G
L X H Y T F G N B E I S E F L
P Y J T D H S E I D N S G B A
I M P H E R O M O N E B G F N
N O R E P I N E P H R I N E D
S G W Y J T U D H R T K F I
P T Y E R W Q A S E D F R G N
```

calcitonin
cortisol
cretinism
hormone
insulin
myxedema
norepinephrine
oxytocin
pheromone
prostaglandin

a. _Calcitonin_ Thyroid hormone that regulates blood calcium.
b. _Hormone_ Chemical messenger.
c. _Cretinism_ Condition due to improper development of thyroid in infants.
d. _Oxytocin_ Posterior pituitary hormone causing uterine contractions and milk letdown.
e. _insulin_ Pancreatic hormone that lowers blood glucose.
f. _cortisol_ Adrenal gland hormone that increases blood glucose.
g. _myxedema_ Chemical substance that influences behavior of mate.
h. _myxedema_ Condition caused by lack of thyroid hormone in adult.
i. _norepinephrine_ Stress hormone from adrenal medulla.
j. _____ Local tissue hormone.

CHAPTER TEST

OBJECTIVE TEST

Do not refer to the text when taking this test.

_a_ 1. All hormones are believed to
   a. have membrane receptors.
   b. affect cellular metabolism.
   c. increase the amount of cAMP.
   d. increase the amount of protein synthesis.

_d_ 2. The adrenal glands are
   a. at the base of the brain.
   b. on the trachea.
   c. on the kidney.
   d. beneath the stomach.

_a_ 3. Which statement is *not* true about hormones?
   a. Hormones search throughout the bloodstream for their receptors.
   b. They act as chemical messengers.
   c. They are released by endocrine glands.
   d. They can affect our appearance, our metabolism, or our behavior.

_a_ 4. Hormonal secretions are often controlled by
   a. negative feedback mechanisms.
   b. positive feedback mechanisms.
   c. the hormone insulin.
   d. the cerebrum of the brain.

_b_ 5. Steroid hormones
   a. combine with hormone receptors in the plasma membrane.
   b. pass through the membrane.
   c. activate genes leading to protein synthesis.
   d. Both *b* and *c* are correct.

_b_ 6. Which gland produces the greatest number of hormones?
   a. posterior pituitary
   b. anterior pituitary
   c. thymus
   d. pineal gland

124

d 7. The hypothalamus controls the anterior pituitary via
   a. nervous stimulation.
   b. the midbrain.
   c. vasopressin.
   d. releasing hormones.

b 8. ADH and oxytocin are
   a. produced by the hypothalamus.
   b. secreted by the posterior pituitary.
   c. secreted by the thyroid gland.
   d. secreted by the parathyroids.

b 9. Which hormone is involved with milk production and nursing?
   a. prolactin
   b. oxytocin
   c. ADH
   d. GH

e 10. The anterior pituitary stimulates the
   a. thyroid.
   b. adrenal cortex.
   c. adrenal medulla.
   d. pancreas.
   e. Both *a* and *b* are correct.

____ 11. Too much urine matches too
   a. little ADH.
   b. much ADH.
   c. little ACTH.
   d. much ACTH.

d 12. Thyroxin
   a. increases metabolism.
   b. stimulates the thyroid gland.
   c. lowers oxygen uptake.
   d. All of these are correct.

q 13. The adrenal cortex produces hormones affecting
   a. glucose metabolism.
   b. amino acid metabolism.
   c. sodium balance.
   d. All of these are correct.

a 14. Which hormone regulates blood calcium levels?
   a. calcitonin
   b. parathyroid hormone
   c. cortisol
   d. Both *a* and *b* are correct.

c 15. Which gland produces sex hormones?
   a. anterior pituitary
   b. posterior pituitary
   c. adrenal cortex
   d. Both *a* and *b* are correct.

v 16. Tetany occurs when there is too
   a. little calcium in the blood.
   b. much calcium in the blood.
   c. little sodium in the blood.
   d. much sodium in the blood.

q 17. Cushing syndrome is due to a malfunctioning
   a. thyroid.
   b. adrenal cortex.
   c. adrenal medulla.
   d. pancreas.

b 18. A simple goiter is caused by
   a. too much salt in the diet.
   b. too little iodine in the diet.
   c. too many sweets in the diet.
   d. a bland diet.

____ 19. Acromegaly might be due to a tumor of the
   a. pancreas.
   b. anterior pituitary.
   c. thyroid.
   d. adrenal cortex.

a 20. If a person is suffering from insulin shock, he or she should
   a. be given some sugar.
   b. sit with the head down.
   c. be given insulin.
   d. not eat fatty foods.

____ 21. Diabetes insipidus is a disease of the
   a. pancreas.
   b. adrenal cortex.
   c. posterior pituitary.
   d. Both *a* and *b* are correct.

____ 22. In which case is insulin not produced?
   a. type I diabetes
   b. type II diabetes
   c. type III diabetes
   d. diabetes insipidus

____ 23. One of the chief differences between pheromones and local hormones is
   a. one is a chemical messenger and the other is not.
   b. the distance over which they act.
   c. one is made by invertebrates and the other by vertebrates.
   d. All of these are correct.

a 24. The endocrine system produces epinephrine, which increases blood pressure. The system most directly affected by the secretion of epinephrine for blood pressure adjustments is the
   a. respiratory system.
   b. circulatory system.
   c. urinary system.
   d. reproductive system.

c 25. The gonads that produce sex hormones technically belong to which system?
   a. lymphatic system
   b. nervous system
   c. reproductive system
   d. urinary system

Thought Questions

Answer in complete sentences.

26. Explain the occurrence of a goiter when an individual does not receive enough iodine in the diet.

27. Why does the release of renin by the kidneys cause the blood pressure to rise?

Test Results: _____ number correct ÷ 27 = _____ × 100 = _____%

Answer Key

Study Questions

1. a. pheromones **b.** insulin (any endocrine gland hormone) **c.** neurotransmitter **2.** b **3. a.** peptide hormone **b.** hormone receptor **c.** cyclic AMP **d.** active **e.** steroid hormone **f.** hormone receptor complex **g.** protein synthesis **4. a.** hypothalamus, hypothalamic-releasing hormone, ADH **b.** pituitary gland, growth hormone, ATCH **c.** pineal gland, melatonin **d.** thyroid gland, thyroxin, calcitonin **e.** parathyroid hormone **f.** thymus, thymosin **g.** adrenal gland, cortisol, aldosterone, epinephrine, norepinephrine **h.** pancreas, insulin, glucagon **i.** ovary, estrogen, progesterone **j.** testis, testosterone **5. a.** decrease **b.** negative **c.** contrary **6.** Dotted lines should be drawn between positions (1) and (2) and between (2) and (3) to indicate negative feedback. **7. a.** TRH (thyroid-releasing hormone) **b.** TSH (thyroid-stimulating hormone) **c.** thyroxine **d.** ACRH (adrenocorticoid-releasing hormone) **e.** ACTH (adrenocorticotropic hormone) **f.** cortisol **8. a.** PP **b.** AP **c.** PP **d.** AP **9. a.** thyroid **b.** thyroxin **c.** adrenal cortex **d.** cortisol **e.** ovaries **f.** estrogen, progesterone **g.** testes **h.** testosterone **10.** negative feedback control **11. a.** prolactin **b.** Melanocyte-stimulating **c.** skeletal **d.** dwarf **e.** giant **f.** acromegaly **g.** enlarge **12. a.** cretinism, simple goiter, myxedema **b.** exophthalmic goiter **c.** cretinism **d.** myxedema **e.** simple goiter **13. a.** Ca^{2+} in bones **b.** Ca^{2+} in bones **c.** Ca^{2+} in blood **d.** Ca^{2+} in blood **e.** Ca^{2+} in blood **f.** Ca^{2+} in bones **14. a.** norepinephrine and epinephrine prepare the body for emergencies **b.** nervous impulse **15. a.** yes, no **b.** yes, no **c.** no, yes **d.** yes, no **e.** no, yes **f.** yes, no **g.** no, yes **16. a.** AD **b.** AD **c.** CS **d.** AD **e.** CS **f.** CS **g.** AD **h.** CS **17. a.** insulin **b.** glucagon **18. a.** increases **b.** decreases **c.** decreases **19. a.** melatonin **b.** sleep **c.** estrogen **d.** progesterone **e.** testosterone **f.** sperm **20. a.** exophthalmic goiter **b.** Addison disease **c.** osteoporosis **d.** acromegaly **e.** diabetes mellitus **f.** exophthalmic goiter **g.** Cushing disease and Addison disease **h.** diabetes mellitus **i.** diabetes insipidus **21.** i, a, j, b, c, h, e, g, f, d

Hormone Hockey

First goal: **1.** f **2.** i **3.** h **4.** d **5.** b. Second goal: **6.** f **7.** b **8.** d **9.** d **10.** a. Third goal: **11.** i **12.** h **13.** b **14.** a **15.** l. Fourth goal: **16.** c **17.** g and h **18.** d and e **19.** j **20.** f. Fifth goal: **21.** yes **22.** no **23.** yes **24.** no **25.** no **26.** yes **27.** yes **28.** no **29.** no **30.** yes

DEFINITIONS WORDSEARCH

```
    C R E T I N I S M           P
  O A   H                       R
  R A L   O X Y T O C I N       O
  T M   C   R           N       S
  I E   I   M           S       T
  S D   T   O           U       A
  O E   O       N       L       G
  L X           N   E   I       L
  Y             I       N       A
    M P H E R O M O N E         N
  N O R E P I N E P H R I N E D
                                I
                                N
```

a. calcitonin   **b.** hormone   **c.** cretinism   **d.** oxytocin
e. insulin   **f.** cortisol   **g.** pheromone   **h.** myxedema
i. norepinephrine   **j.** prostaglandin

CHAPTER TEST

1. b **2.** c **3.** a **4.** a **5.** d **6.** b **7.** d **8.** b **9.** b
10. e **11.** a **12.** a **13.** d **14.** d **15.** c **16.** a
17. b **18.** b **19.** b **20.** a **21.** c **22.** a **23.** b
24. b **25.** c **26.** When an individual does not receive enough iodine in the diet, the thyroid is unable to produce thyroxin. The lack of thyroxin in the blood causes the anterior pituitary to produce more TSH, and this hormone promotes increase in the size of the thyroid.
27. Renin leads to the formation of angiotensin II, which stimulates the adrenal cortex to secrete aldosterone. Aldosterone causes sodium to be reabsorbed by the kidneys, and this leads to an increase in blood volume and blood pressure.

PART 5 REPRODUCTION IN HUMANS

15

REPRODUCTIVE SYSTEM

STUDY TIPS

It is by virtue of the human reproductive system that we propagate our species. In fact, the reproductive system does not participate in maintaining homeostasis to the extent that other bodily systems do. Its hormones help us to maintain health and vigor but are not a necessity toward that end.

Study the diagrams of the male (p. 318) and female (p. 322) reproductive tracts. Know the functions of each structure and how they participate in the reproductive process. Reproduction is under hormonal control, beginning with the hypothalamus. It triggers the release of FSH and LH from the anterior pituitary in both males and females. These two hormones then influence the production of gametes as well as the production of the male and female hormones responsible for secondary sexual characteristics. Make a chart of the ovarian cycle of female hormones that control the uterine cycle during the monthly menstrual cycle (pp. 325–327).

This chapter closes with a discussion of methods of conception control (pp. 330–333). Study this information carefully. The method of birth control you choose may also be able to protect you from acquiring many sexually transmitted diseases, the subject of the next chapter.

STUDY QUESTIONS

Study the text section by section as you answer the questions that follow.

15.1 MALE REPRODUCTIVE SYSTEM (PAGES 318–321)

- Describe the structure and function of the male reproductive system.
- State the path sperm take from the site of production until they exit the male.
- Name the glands that add secretions to seminal fluid.
- Discuss hormonal regulation of sperm production in the male.
- Name the actions of testosterone, including both primary and secondary sex characteristics.

1. Using the alphabetized list of terms and blanks provided, identify and state a function for the parts of the human male reproductive system shown in the following diagram.

    bulbourethral gland, epididymis, penis, prostate gland, seminal vesicles, testis, urethra, urinary bladder, vas deferens

    Structure   **Function**

    a. _____   _____
    b. _____   _____
    c. _____   _____
    d. _____   _____
    e. _____   _____
    f. _____   _____
    g. _____   _____
    h. _____   _____
    i. _____   _____

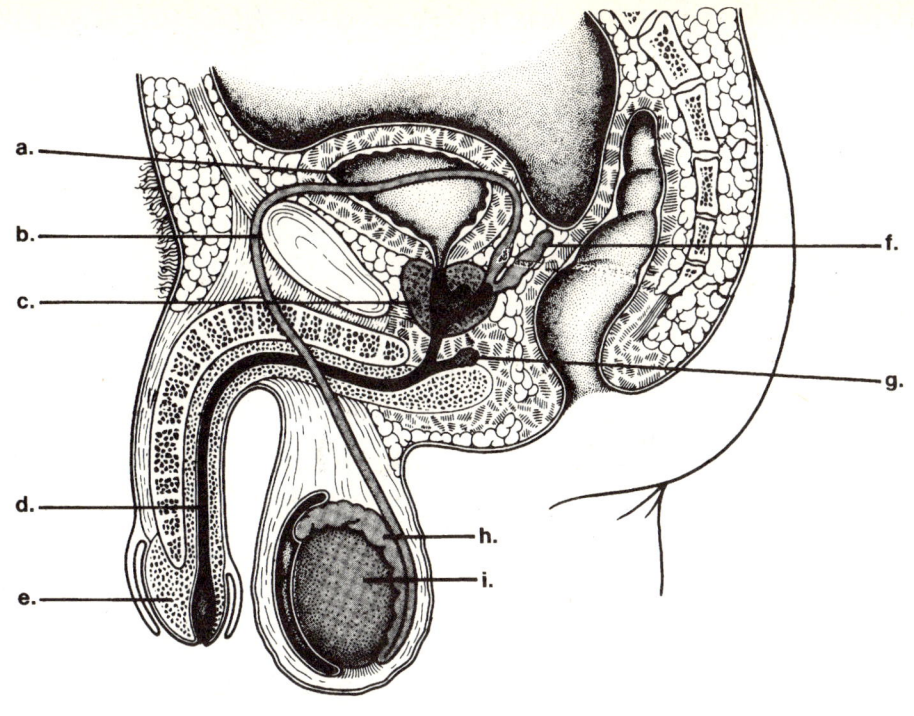

2. Write *ST* beside the statements that apply to the seminiferous tubules and *IC* beside the statements that apply to the interstitial cells.
   _____ a. produce androgens
   _____ b. produce sperm
   _____ c. controlled by FSH
   _____ d. controlled by LH

3. Trace the path of sperm through the male reproductive system.

   Testes to the a._____ to the vas deferens to the b._____.

4. What three organs add secretions to seminal fluid?
   a._____

   b._____

   c._____

5. What is the general function of these secretions? _____
   _____

6. The process of sperm production is called a._____. This occurs inside
   b._____ tubules inside each testis. Helper cells, known as c._____ cells,
   nourish and regulate the developing sperm cells.

7. Mature sperm cells have three parts: a._____, b._____ and c._____.
   There is also an enzyme-containing d._____ at the end to allow entrance into the cell layers
   surrounding the ovum. What section of a sperm cell contains the mitochondria that provide energy for
   motility? e._____

8. Indicate whether the following statements are true (T) or false (F). Rewrite the false statements to make true statements.

   a._____ Testosterone exerts negative feedback control over the anterior pituitary secretion of LH. Rewrite: __
   _____

   b._____ Inhibin exerts negative feedback control over the anterior pituitary secretion of FSH. Rewrite: _____
   _____

129

9. What are some of the effects of testosterone on the development of secondary sexual characteristics?
   _____
   _____

15.2 FEMALE REPRODUCTIVE SYSTEM (PAGES 322–323)

- Describe the structure and function of the female reproductive system.
- Describe the structure of the ovaries.
- Label a diagram of the external female genitalia.

10. Using the alphabetized list of terms and the blanks provided, identify and state a function for the parts of the human female reproductive structures and those urinary structures in the following diagram.
    cervix, ovary, oviduct, urethra, urinary bladder, uterus, vagina

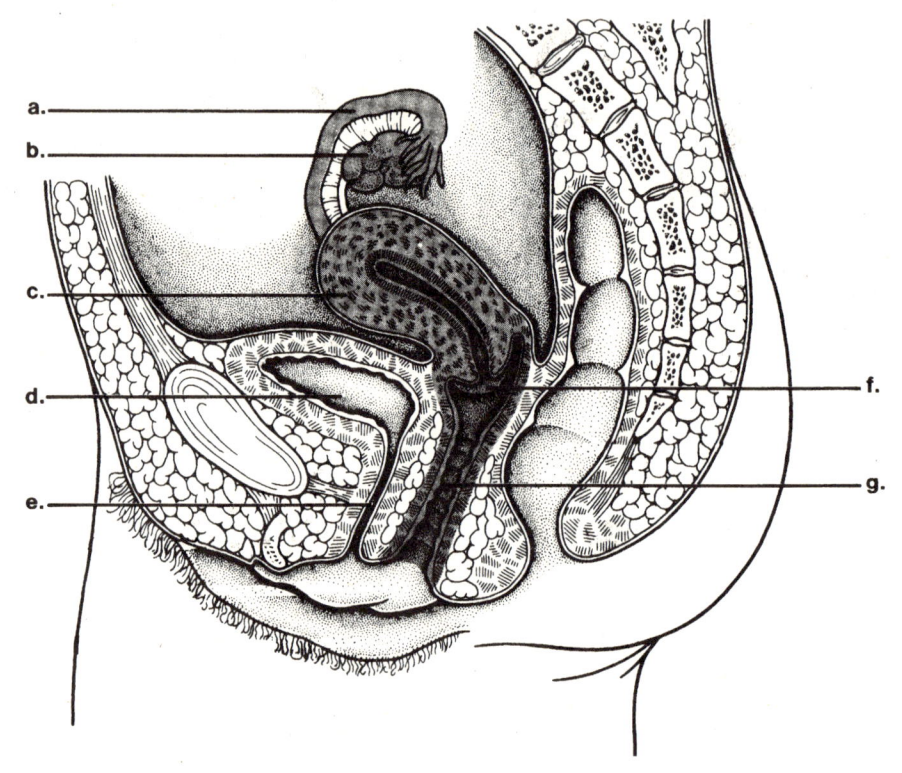

Structure   **Function**

a. _____   _____
b. _____   _____
c. _____   _____
d. _____   _____
e. _____   _____
f. _____   _____
g. _____   _____

11. When sperm enter the female reproductive tract, they are deposited into the a._____. From there, they pass through the b._____ of the uterus. They swim up through the c._____ until they reach the egg cell.

130

12. Label the following diagram of the vulva, using the alphabetized list of terms.

    anus
    glans clitoris
    labia majora
    labia minora
    mons pubis
    urethra
    vagina

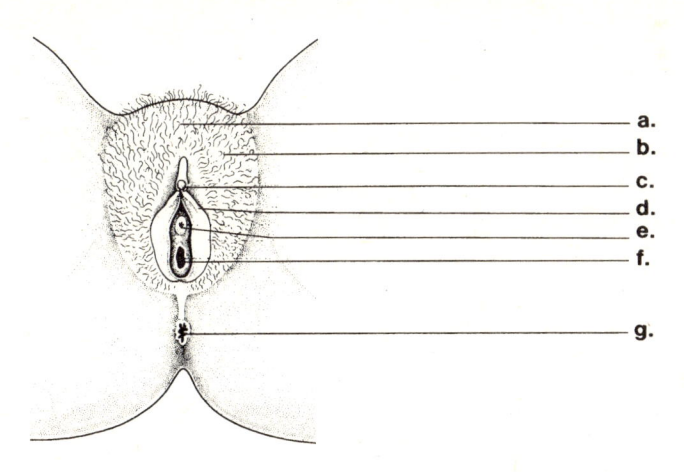

    a. _____
    b. _____
    c. _____
    d. _____
    e. _____
    f. _____
    g. _____

15.3 FEMALE HORMONE LEVELS (PAGES 325–328)

- Describe the ovarian and uterine cycles.
- Discuss hormonal regulation in the female, including feedback control.
- Name the actions of estrogen and progesterone, including the influence on secondary sexual characteristics.

13. Match the items in the key to the phrases that follow.
    Key *estrogen   FSH   LH   progesterone*

    _____ a. gonadotropic hormones

    _____ b. female sex hormones

    _____ c. primarily secreted by follicle

    _____ d. primarily secreted by corpus luteum

14. Each a._____ in the ovary contains an oocyte egg. A secondary follicle develops into a(n) b._____ follicle. c._____ is the release of the secondary oocyte (egg) from the ovary. Following ovulation, a follicle becomes a(n) d._____.

15. What are some of the effects of estrogen on the development of secondary sexual characteristics?
    _____
    _____

16. Ovarian and uterine cycles (simplified, and assuming a 28-day cycle). Fill in the table to indicate the events in these two cycles.

Ant. Pit.	Ovarian Cycle Phases	Events	Uterine Cycle Phases	Events
FSH	Follicular (days 1–13)	a.	b. (days 1–5)	c.
			d. (days 6–13)	e.
	Ovulation (day 14)			
LH	Luteal (days 15–28)	f.	g. (days 15–28)	h.

131

17. Indicate whether the following statements are true (T) or false (F). Rewrite the false statements to make true statements.

    _____ a. Pregnancy occurs as soon as fertilization occurs. Rewrite: _____
    _____

    _____ b. HCG prevents degeneration of the corpus luteum. Rewrite: _____
    _____

    _____ c. During pregnancy, ovulation continues because estrogen and progesterone are still present. Rewrite:
    _____

15.4 DEVELOPMENT OF MALE AND FEMALE SEX ORGANS (PAGE 328)

- Compare the development of gonads in female and male fetuses.

18. It is the genes on the a._____ chromosome that determine whether the embryo will become male or female. Gonads begin developing internally in the b._____ week of gestation. At six weeks, a small bud appears between the legs of the embryo. It will develop into a(n) c._____ if it is a female and into a(n) d._____ if it is a male.

15.5 CONTROL OF REPRODUCTION (PAGES 330–334)

- Categorize birth-control measures by the criteria used in the text.
- List the causes of infertility and the various alternative methods of reproduction.

19. Following are two groups of birth-control measures. Rank the members of each group from the best (1) to the worst (4) on the basis of effectiveness.

    A
    _____ a. coitus interruptus
    _____ b. spermicidal jelly/cream
    _____ c. condom + spermicide
    _____ d. natural family planning

    B
    _____ e. vasectomy
    _____ f. natural family planning
    _____ g. diaphragm + spermicide
    _____ h. IUD

20. The two common causes of infertility in females are a._____ caused by b._____ and c._____ caused by d._____.

21. The most common cause of infertility in males is a._____ caused by b._____.

22. In which alternative reproductive method is the egg fertilized in laboratory glassware? _____
    _____

15.6 HOMEOSTASIS (PAGE 334)

- Discuss how the reproductive system works with other systems of the body to maintain homeostasis.

23. The *Working Together* diagram in your textbook shows how the reproductive system works with the other organ systems of the body. Match the organ systems with the correct descriptions.
    a. *Sex hormones influence cardiovascular health; sexual activities stimulate circulatory system.*
    b. *Sexual activity increases breathing; pregnancy causes breathing rate and vital capacity to increase.*
    c. *Sex hormones masculinize or feminize the brain, exert feedback control over the hypothalamus, and influence sexual behavior.*
    d. *Gonads produce the sex hormones.*
    e. *Androgens activate oil glands; sex hormones stimulate fat deposition, affect hair distribution in males and females.*
    f. *Penis in males contains the urethra and performs urination; prostate enlargement hinders urination.*
    g. *Androgens promote growth of skeletal muscle.*
    h. *Sex hormones influence immune functioning; acidity of vagina helps prevent pathogen invasion of body; milk passes antibodies to newborn.*
    i. *Pregnancy crowds digestive organs and promotes heartburn and constipation.*
    j. *Sex hormones influence bone growth and density in males and females.*

_____ integumentary system
_____ lymphatic system
_____ circulatory system
_____ skeletal system
_____ urinary system
_____ muscular system
_____ respiratory system
_____ reproductive system
_____ endocrine system
_____ digestive system

DEFINITIONS WORDSEARCH

Review key terms by completing this wordsearch, using the following alphabetized list of terms:

```
E T A T S O R P A P S M E A R
N B V C X X S A M U T O R C S
T E S T R O G E N T X V V V I
L K J H T R E W Q E C A V B T
P O I C E R V I X R F R C A S
B E S T A V W S X U H Y Z S E
V F I M B R I A S S A D F B T
G Y E F S M E N O P A U S E L
```

cervix
estrogen
fimbria
menopause
ovary
Pap smear
prostate
scrotum
testis
uterus

a. _____ Narrow base of the uterus leading to vagina.
b. _____ Female sex hormone responsible for secondary sexual characteristics.
c. _____ Muscular organ in which fetus develops.
d. _____ Pouch of skin that encloses testes.
e. _____ Fingerlike extensions of oviduct.
f. _____ Termination of menstrual cycle in older women.
g. _____ Organ that produces sperm.
h. _____ Organ that produces eggs.
i. _____ Doughnut-shaped gland around male urethra.
j. _____ Clinical test to detect cervical cancer.

CHAPTER TEST

OBJECTIVE TEST

Do not refer to the text when taking this test.

____ 1. The vas deferens
 a. becomes erect.
 b. carries sperm.
 c. is surrounded by the prostate gland.
 d. runs through bulbourethral glands.

____ 2. The prostate gland
 a. is removed when a vasectomy is performed.
 b. is not needed to maintain the secondary sex characteristics.
 c. receives urine from the bladder.
 d. almost never becomes cancerous.

____ 3. Which gland or organ secretes hormones?
 a. seminal vesicles
 b. prostate gland
 c. bulbourethral gland
 d. testes

____ 4. FSH
 a. stimulates sperm production in males.
 b. stimulates development of the follicle in females.
 c. is produced by the anterior pituitary.
 d. All of these are correct.

____ 5. Gonadotropic hormones are produced by the
 a. testes.
 b. ovaries.
 c. anterior pituitary.
 d. uterus.

___ 6. Which hormone stimulates the production of testosterone?
   a. LH
   b. FSH
   c. estrogen
   d. inhibin

___ 7. Which hormone regulates the production of testosterone?
   a. LH
   b. FSH
   c. estrogen
   d. inhibin

___ 8. The urethra is part of the reproductive tract in
   a. the female.
   b. the male.
   c. both the male and female.
   d. invertebrates.

___ 9. Which chromosome contains genes that determine whether a developing embryo develops into a female or male?
   a. X chromosome
   b. Y chromosome
   c. chromosome #3
   d. chromosome #21

___ 10. The endometrium
   a. lines the vagina.
   b. is involved in the formation of the placenta.
   c. produces estrogen.
   d. None of these is correct.

___ 11. The uterus
   a. is connected to both the oviducts and vagina.
   b. is not an endocrine gland.
   c. contributes to the development of the placenta.
   d. All of these are correct.

___ 12. Which structure is present after ovulation?
   a. primary follicle
   b. secondary follicle
   c. Graafian follicle
   d. corpus luteum

___ 13. Ovulation occurs
   a. due to hormonal changes.
   b. always on day 14.
   c. in postmenopausal women.
   d. as a result of intercourse.

___ 14. Which of these secretes hormones involved in the ovarian cycle?
   a. hypothalamus
   b. anterior pituitary gland
   c. ovary
   d. All of these are correct.

___ 15. FSH stimulates the
   a. release of an egg cell from the follicle.
   b. development of a follicle.
   c. development of the endometrium.
   d. beginning of menstrual flow.

___ 16. Secretions from which of the following structures are required before implantation can occur?
   a. the ovarian follicle
   b. the pituitary gland
   c. the corpus luteum
   d. All of these are correct.

___ 17. Human chorionic gonadotropin (HCG) is different from other gonadotropic hormones because it
   a. is produced by the maternal portion of the placenta.
   b. is not produced by a female endocrine gland.
   c. does not stimulate any tissue in the body.
   d. does not enter the bloodstream.

___ 18. Pregnancy begins
   a. when an egg is fertilized.
   b. when ovulation occurs.
   c. upon successful implantation.
   d. during the follicular phase.

___ 19. Menstruation begins in response to
   a. increasing estrogen levels.
   b. decreasing progesterone levels.
   c. changes in blood chemistry.
   d. secretion of FSH.

___ 20. What do all the birth-control methods have in common?
   a. They all use some device.
   b. They all interrupt intercourse.
   c. They are all terribly expensive and uncomfortable.
   d. None of these is correct.

___ 21. A vasectomy
   a. prevents the egg from reaching the oviduct.
   b. prevents sperm from reaching seminal fluid.
   c. prevents release of seminal fluid.
   d. inhibits sperm production.

___ 22. Which of these means of birth control prevents implantation?
   a. diaphragm
   b. IUD
   c. cervical cap
   d. vaginal sponge

___ 23. In vitro fertilization occurs in
   a. the vagina.
   b. a surrogate mother.
   c. laboratory glassware.
   d. the uterus.

___ 24. The _____ system provides nutrients for the growth of a developing fetus.
   a. digestive
   b. respiratory
   c. muscular
   d. lymphatic

___ 25. Androgens from the reproductive system stimulate the growth of skeletal muscle from which system?
   a. digestive system
   b. muscular system
   c. urinary system
   d. respiratory system

THOUGHT QUESTIONS

Answer in complete sentences.

26. How does the structure of a sperm suit its function?

27. Why do you expect to find sex hormones from the ovaries in pregnant women but not in menopausal women?

Test Results: _____ number correct ÷ 27 = _____ × 100 = _____%

ANSWER KEY

STUDY QUESTIONS

1. a. urinary bladder; stores urine **b.** vas deferens; conducts and stores sperm **c.** prostate gland; contributes to seminal fluid **d.** urethra; conducts both urine and sperm **e.** penis; organ of sexual intercourse **f.** seminal vesicles; contribute to seminal fluid **g.** bulbourethral gland; contributes nutrients and fluid to semen **h.** epididymis; stores sperm as they mature **i.** testis; production of sperm and male sex hormones **2. a.** IC **b.** ST **c.** ST **d.** IC **3. a.** epididymis **b.** urethra **4. a.** seminal vesicles **b.** prostate gland **c.** bulbourethral glands **5.** To nourish sperm cells, to increase the motility of sperm cells, and for lubrication. **6. a.** spermatogenesis **b.** seminiferous **c.** sustentacular (Sertoli) **7. a.** head **b.** midpiece **c.** tail **d.** acrosome **e.** midpiece **8. a.** T **b.** T **9.** Testosterone deepens the voice, promotes the development of muscles and body and facial hair, increases secretions from oil glands, and promotes the development of the sexual organs. **10. a.** oviduct; conduction of egg **b.** ovary; production of eggs and sex hormones **c.** uterus; houses developing fetus **d.** urinary bladder; storage of urine **e.** urethra; conduction of urine **f.** cervix; opening of uterus **g.** vagina; receives penis during sexual intercourse and serves as birth canal **11. a.** vagina **b.** cervix **c.** oviduct **12. a.** mons pubis **b.** labia majora **c.** glans clitoris **d.** labia minora **e.** urethra **f.** vagina **g.** anus **13. a.** FSH and LH **b.** progesterone and estrogen **c.** estrogen **d.** progesterone **14. a.** follicle **b.** Graafian **c.** Ovulation **d.** corpus luteum **15.** Estrogen promotes the deposition of body fat, the maturation and maintenance of the sexual organs, and breast development. **16. a.** follicle matures and produces egg and estrogen **b.** menstruation **c.** endometrium breaks down **d.** proliferative **e.** endometrium rebuilds **f.** corpus luteum produces progesterone **g.** secretory **h.** endometrium thickens and glands are secretory **17. a.** F; Pregnancy occurs as soon as implantation occurs. **b.** T **c.** F. During pregnancy, ovulation discontinues because estrogen and progesterone secreted by the corpus luteum and the placenta exert feedback control over the hypothalamus and the anterior pituitary. **18. a.** Y **b.** 7th **c.** clitoris **d.** penis **19. a.** 3 **b.** 2 **c.** 1 **d.** 4 **e.** 1 **f.** 4 **g.** 3 **h.** 2 **20. a.** blocked oviducts **b.** PID **c.** failure to ovulate **d.** low body weight **21. a.** low sperm count or large proportion of abnormal sperm **b.** environmental factors **22.** in vitro fertilization **23.** e, h, a, j, f, g, b, d, c, i

DEFINITIONS WORDSEARCH

```
E T A T S O R P A P S M E A R
                M U T O R C S
E S T R O G E N T   V       I
                E   A       T
      C E R V I X R R       S
                U   Y       E
F I M B R I A   S           T
      M E N O P A U S E
```

a. cervix **b.** estrogen **c.** uterus **d.** scrotum **e.** fimbria **f.** menopause **g.** testis **h.** ovary **i.** prostate **j.** Pap smear

CHAPTER TEST

1. b **2.** b **3.** d **4.** d **5.** c **6.** a **7.** d **8.** b **9.** b **10.** b **11.** d **12.** d **13.** a **14.** d **15.** b **16.** d **17.** b **18.** c **19.** b **20.** d **21.** b **22.** b **23.** c **24.** a **25.** b **26.** A sperm functions to fertilize an egg, and its various parts are specialized; for example, the head is capped by the acrosome, which releases enzymes that allow the sperm to penetrate the egg; the head contains 23 chromosomes; the midpiece contains mitochondria that provide energy; and the tail is a flagellum that allows the sperm cell to swim. **27.** The ovaries do not secrete female sex hormones in pregnant women or in menopausal women. During pregnancy, however, ovarian hormones are replaced by hormones secreted by the placenta. The placental hormones help to maintain the uterine lining, and thus, the pregnancy.

16

SEXUALLY TRANSMITTED DISEASES

STUDY TIPS

No one expects to contract sexually transmitted diseases (STDs), yet every year, millions of people do. Many STDs might be considered annoyances, but others can lead to sterility through pelvic inflammatory complications in women (p. 348), cancer (p. 342), or even death. Everyone should fully understand the preventative measures (p. 339) to avoid contracting or transmitting STDs.

STDs are caused by a variety of pathogens (p. 340), including viruses (pp. 340-345), bacteria (p. 346-350), and others, such as pubic lice (p. 351). Among the most interesting are the viruses and retroviruses, because of the manner in which they alter our DNA to make more copies of themselves. The AIDS virus is a retrovirus that invades helper T lymphocytes (p. 342). Make a diagram detailing how viruses enter cells and make alterations in our DNA. Retroviruses go about the same process in a different manner; remember to note the differences.

Compile a chart listing the STDs discussed in this chapter, the pathogens causing them, symptoms of the disease, dangerous side effects of infection, and whether or not the disease can be cured. You might also find it interesting to include a column indicating the estimated number of new cases of the disease each year.

STUDY QUESTIONS

Study the text section by section as you answer the questions that follow.

16.1 VIRAL IN ORIGIN (PAGES 340–345)

- Describe the structure and life cycle of a DNA and an RNA virus.
- Discuss the three stages of AIDS, how the disease can be treated, and how to prevent its occurrence.
- Name the causative agent for AIDS, genital herpes, and genital warts.
- Describe the symptoms of a herpes infection.
- Distinguish between strains of hepatitis, and know which is transmitted through sexual contact.

1. What are the three major parts of a virus?
   a. _____ b. _____ c. _____

2. After a virus enters a cell, what two events are needed before the assemblage of new DNA viruses can take place?
   a. _____ b. _____

3. How does the life cycle of a retrovirus differ from that of a DNA virus? _____

4. What name is given to the retrovirus that causes AIDS? a._____ What specific type of T lymphocyte is the host for the AIDS retrovirus? b._____

5. Describe the symptoms of the following stages of an HIV infection:
   Category A a. _____

   Category B b. _____

136

Category C c. _____
_____

6. How is the use of intravenous drugs related to acquiring HIV? _____
_____

7. Highly active antiretroviral therapy (HAART) uses two drugs that inhibit a. _____ and one that inhibits protease needed for b. _____. The experimental vaccines use only a portion of the c. _____ to stimulate the production of d. _____ by the body.

8. What virus causes genital warts? a. _____ What other human disease is caused by this virus? b. _____

9. What virus causes genital herpes? a. _____ What are the symptoms of genital herpes? b. _____
Where does the virus reside between outbreaks? c. _____ What problems does this disease cause in newborns when they contract it from their mothers during the birth process? d. _____

10. Which form of hepatitis is the one spread mostly by sexual contact? a. _____ Is there a vaccine? b. _____
What are the symptoms of this disease? c. _____

16.2 BACTERIAL IN ORIGIN (PAGES 346–350)

- Describe the structure of a prokaryotic cell, the three bacterial shapes, and the life cycle of a bacterium.
- Understand the methods for preventing the spread of infectious bacteria.
- Name the causative agent for chlamydia, gonorrhea, and syphilis.
- Describe the symptoms of chlamydia in the male and in the female, and compare this disease to a gonorrheal infection.
- Describe the symptoms of a gonorrheal infection in the male and in the female.
- Define PID, and describe how it affects reproduction.
- List the three stages of syphilis.
- Describe how a newborn becomes infected with AIDS, herpes, gonorrhea, chlamydia, or syphilis.

11. Label the shapes of bacteria in this drawing.

12. How does a prokaryotic cell differ from a human eukaryotic cell?
    a. _____
    b. _____
    c. _____

13. Most bacteria are free-living decomposers that get their nutrients by _____
_____

137

14. Bacterial reproduction:
    What type of reproduction is typical of bacteria? a. _____

    What is an endospore, and what is its function? b. _____
    _____

15. *Chlamydia trachomatis* is the most common cause of _____.

16. Name four ways in which chlamydia is like gonorrhea.
    a. _____   b. _____
    c. _____   d. _____

17. What is the danger of pelvic inflammatory disease (PID)? _____
    _____

18. What organism causes gonorrhea? _____

19. What are the symptoms of gonorrhea? a. _____
    Does it spread to other areas of the body? b. _____

20. What is the cause of syphilis? _____

21. Describe the three stages of syphilis.
    primary   a. _____
    secondary b. _____
    tertiary c. _____

22. Which three sexually transmitted diseases passed on to the newborn could be prevented by caesarean delivery?
    a. _____   b. _____   c. _____

16.3 OTHER SEXUALLY TRANSMITTED DISEASES (PAGE 351)

- List other common sexually transmitted diseases, and describe the associated symptoms.

23. What condition do these organisms cause?
    Trichomonas vaginalis a. _____
    Candida albicans b. _____
    Pubic lice c. _____

DEFINITIONS WORDSEARCH

Review key terms by completing this wordsearch, using the following alphabetized list of terms:

```
G D F R E D U T H E D O P
E D G W T I S D E V N U A
N B O P I P A T H O G E N
I J A U Y T G A R E D N T
T R E T R O V I R U S D I
A G U Y P T R D E W E O B
L T U H J R H S F D J S I
W G H M D S R V C F R P O
A D E S M O I U N E S O T
R F D R V A D W Z S A R I
T H E P A T I T I S Y E C
S U B A C T E R I U M A C
```

AIDS
antibiotic
bacterium
endospore
genital warts
gumma
hepatitis
pathogen
PID
retrovirus

a. _____ Large, unpleasant sore appearing during the tertiary stage of syphilis.
b. _____ Inflammation of the liver.
c. _____ Virus containing RNA and carrying out reverse transcription.
d. _____ Simple unicellular organism.
e. _____ Disease-causing agent.
f. _____ Resistant stage of a bacterium.
g. _____ Medicine that interferes with bacterial metabolism.
h. _____ Human papillomavirus causes these.
i. _____ Acquired immunodeficiency syndrome.
j. _____ Pelvic inflammatory disease.

Chapter Test

Objective Test

Do not refer to the text when taking this test.

____ 1. A retrovirus is
  a. the cause of herpes.
  b. a DNA virus.
  c. capable of reverse transcription.
  d. Both *b* and *c* are correct.

____ 2. Which of these is characteristic of full-blown AIDS?
  a. pneumonia
  b. Kaposi's sarcoma
  c. psychological disturbances
  d. All of these are correct.

____ 3. The cause of AIDS is a
  a. retrovirus.
  b. virus.
  c. bacterium.
  d. protozoan.

____ 4. Which virus causes genital warts?
  a. herpes simplex type 2
  b. human papillomavirus
  c. tobacco mosaic virus
  d. hepatitis B virus

____ 5. Which form of hepatitis is the one contracted sexually?
  a. hepatitis A
  b. hepatitis B
  c. hepatitis C
  d. None are contracted sexually.

____ 6. Which virus causes genital herpes?
  a. herpes simplex type 2
  b. human papillomavirus
  c. tobacco mosaic virus
  d. hepatitis B virus

____ 7. When an infant comes in contact with herpes lesions in his/her mother's birth canal, what complications can be expected?
  a. possible brain damage
  b. possible blindness
  c. possible death
  d. All of these are correct.

____ 8. The name of the organism that causes syphilis is
  a. *Treponema pallidum.*
  b. HIV.
  c. *Neisseria gonorrhoeae.*
  d. *Trichomonas vaginalis.*

____ 9. Congenital gonorrhea is caused by
  a. bacteria crossing the placenta.
  b. bacteria in the vagina.
  c. bacteria in the oviducts.
  d. All of these are correct.

____ 10. Which phrase best describes most bacteria?
  a. obligate parasite
  b. reproduction inside host cell
  c. decomposers
  d. Both *a* and *c* are correct.

____ 11. Which of these is a cell?
  a. bacteria
  b. virus
  c. Both of these are correct.
  d. Neither of these is correct.

____ 12. Which has a cell with a nucleus?
  a. bacteria
  b. virus
  c. Both of these are correct.
  d. Neither of these is correct.

____ 13. Decomposing bacteria
  a. utilize a host.
  b. always cause a disease.
  c. are free living.
  d. Both *a* and *b* are correct.

____ 14. Antibiotics cure
  a. viral infections.
  b. bacterial infections.
  c. fungal infections.
  d. protozoan infections.

139

_____ 15. You can cure gonorrhea by
   a. refraining from sexual intercourse.
   b. taking warm baths.
   c. following antibiotic therapy.
   d. taking AZT.
_____ 16. Chlamydia is similar to
   a. syphilis.
   b. gonorrhea.
   c. genital herpes.
   d. genital warts.
_____ 17. In which stage of syphilis does the victim break out in a rash?
   a. first
   b. second
   c. third
   d. fourth
_____ 18. The main complication from prolonged syphilis is
   a. PID.
   b. reduced immunity.
   c. heart and brain involvement.
   d. genital lesions.
_____ 19. Syphilis has _____ stage(s).
   a. one
   b. two
   c. three
   d. four
_____ 20. Caesarean section is helpful to prevent infection of the newborn with
   a. AIDS.
   b. syphilis.
   c. herpes.
   d. All of these are correct.

_____ 21. Newborns get AIDS
   a. as the virus crosses the placenta.
   b. at birth.
   c. from the mother's milk.
   d. , but it is not known how they contract it.
_____ 22. Which of these does *not* cause symptoms of vaginitis?
   a. *Chlamydia*
   b. gonorrhea
   c. syphilis
   d. *Trichomonas*
_____ 23. *Candida albicans*
   a. causes vaginitis.
   b. is a normal organism occurring in the vagina.
   c. is a yeast.
   d. All of these are correct.
_____ 24. Crabs are caused by
   a. a protozoan.
   b. a virus.
   c. pubic lice.
   d. a crab.
_____ 25. Human beings develop active immunity that prevents them from getting sexually transmitted diseases caused by
   a. bacteria.
   b. viruses.
   c. fungi.
   d. None of these is correct.

THOUGHT QUESTIONS

Answer in complete sentences.
26. How does the structure of a virus differ from that of a human cell?

27. Why don't antibiotics cure viral diseases?

Test Results: _____ number correct ÷ 27 = _____ × 100 = _____%

Answer Key

Study Questions

1. a. nucleic acid (DNA or RNA) **b.** capsid **c.** envelope
2. a. replication of DNA **b.** protein synthesis to produce new capsids **3.** Integration of cDNA into the host DNA occurs before transcription to produce viral RNA and translation to produce capsids. **4. a.** HIV (human immunodeficiency virus) **b.** helper T lymphocytes (cells) **5. a.** antibodies in blood, swollen lymph nodes **b.** weight loss, night sweats, fatigue, fever, and diarrhea. **c.** weakness, weight loss, diarrhea, and opportunistic diseases such as Kaposi's sarcoma **6.** Intravenous users often share needles, which directly transfer HIV from one person to the next. **7. a.** reverse transcriptase **b.** viral assembly **c.** envelope or capsid **d.** antibodies **8. a.** human papillomavirus **b.** cervical cancer **9. a.** herpes simplex type 2 virus **b.** blisters that ulcerate, fever, pain upon urination, swollen lymph nodes **c.** in nerve cells near the brain and spinal cord **d.** grave illness, blindness, neurological disorders, brain damage, and death **10. a.** hepatitis B **b.** yes **c.** flu-like symptoms in 50% of those who contract HBV, can be acute or chronic **11. a.** spiral **b.** round **c.** rod **12. a.** lacks organelles **b.** cell wall (capsule perhaps) **c.** moves by flagella **13.** breaking down macromolecules externally, then absorbing the products **14. a.** binary fission **b.** a resistant body formed by bacteria, enabling it to survive during unsuitable environmental conditions **15.** PID (pelvic inflammatory disease) **16. a.** urethritis **b.** PID **c.** cured by antibiotic **d.** affects newborns **17.** it causes sterility in women **18.** *Neisseria gonorrhoeae* **19. a.** most women are asymptomatic until PID develops; men have painful urination and discharge **b.** can spread to internal parts, causing heart damage and arthritis; also can cause eye infections **20.** *Treponema pallidum* **21. a.** chancre **b.** rash **c.** gummas **22. a.** herpes **b.** gonorrhea **c.** chlamydia **23. a.** vaginitis **b.** vaginitis **c.** crabs

Definitions Wordsearch

```
G       D             A
E       I             A
N       A    PATHOGEN N
I       A             T
TRETROVIRUSDI
A       G        D    O B
L       U        S    S I
W       M             P O
A       M             O T
R       A             R I
THEPATITIS  E C
S    BACTERIUM
```

a. gumma **b.** hepatitis **c.** retrovirus **d.** bacterium **e.** pathogen **f.** endospore **g.** antibiotic **h.** genital warts **i.** AIDS **j.** PID.

Chapter Test

1. c **2.** d **3.** a **4.** b **5.** b **6.** a **7.** d **8.** a **9.** b **10.** c **11.** a **12.** d **13.** c **14.** b **15.** c **16.** b **17.** b **18.** c **19.** c **20.** c **21.** a **22.** c **23.** d **24.** c **25.** d **26.** A virus is made up of a capsid and a nucleic acid core, while a human cell has a nucleus and many organelles in a cytoplasm surrounded by a membrane. **27.** Antibiotics interfere with bacterial enzymes not possessed by viruses.

17

DEVELOPMENT AND AGING

STUDY TIPS

Few topics in biology capture the attention of everyone the way human development is able to. We marvel at how two microscopic cells can divide to produce an adult, much less contain all the information to orchestrate each developmental step along the way. In this chapter, you will learn about the stages of development from the embryo to the fetus and about the birthing process. Development from infancy to old age is also discussed.

Make a summary chart (see Table 17.2, p. 374) describing the stages of human development. Include details of early cleavage, morula, and blastocyst stages (p. 366), as well as the later embryonic and fetal stages.

Understanding fetal circulation requires time spent memorizing blood vessels (see Fig. 17.7). Construct a flow diagram indicating the path of circulation from the mother to the fetus, throughout the fetus, and back to the mother.

Pay attention to how the process of aging can be delayed through proper nutrition and exercise (p. 379) and you can benefit from a healthy, long life.

STUDY QUESTIONS

Study the text section by section as you answer the questions that follow.

17.1 FERTILIZATION (PAGES 364–365)

- Describe the process of fertilization, including where and how it occurs.
- Explain the basis of the pregnancy test, and describe a woman's symptoms indicating that she is pregnant.

1. Fertilization usually takes place in the a._____. When sperm make contact with the b._____, they release enzymes from the c._____, and then one sperm can enter the egg. When the sperm nucleus fuses with the egg nucleus, d._____ is complete. As soon as the plasma membrane of the egg and sperm fuse, the e._____ membrane around the egg becomes impenetrable to other sperm. Soon the developing embryo will f._____ itself in the uterine lining and begin producing g._____, the basis of the pregnancy test.

17.2 HUMAN DEVELOPMENT BEFORE BIRTH (PAGES 366–375)

- State the four processes of development.
- Name the extraembryonic membranes, and state the function of each.
- Describe the structure and function of the placenta.
- Trace the path of blood in the fetus.
- Describe the weekly events of embryonic development and the monthly events of fetal development.
- In general, describe the effects of pregnancy on the mother.

2. The process of development: Right after fertilization, the zygote undergoes a series of divisions known as a._____. The solid mass of cells is called a(n) b._____. Next, an interior cavity forms inside the ball of cells, forming a stage called a(n) c._____ with a(n) d._____ off to one side. It is at this stage that implantation occurs. Next, the inner cell mass develops into a(n) e._____. Differentiation continues, and there are f._____ cell layers destined to develop into all the organs and tissues of the body.

3. Development of multicellular organisms consists of a progression of events in which (a) the number of cells increases, (b) the size increases, (c) the form and shape of the individual emerges, and (d) the cells become specialized into various organs. What are the terms used to describe these processes?

   a._____   b._____

   c._____   d._____

4. Using the alphabetized list of terms, label the extraembryonic membranes of the human embryo in the following diagram.

   allantois
   amnion
   chorion
   embryo
   fetal portion of placenta
   maternal portion of placenta
   yolk sac

   Human

   a._____
   b._____
   c._____
   d._____
   e._____
   f._____
   g._____

   umbilical cord

5. Complete this table with the functions of the extraembryonic membranes.

Membrane	Function in Human
Chorion	a.
Amnion	b.
Allantois	c.
Yolk sac	d.

6. The placenta has a(n) a._____ side derived from the extraembryonic membrane, the b._____, and a(n) c._____ side derived from d._____ tissue.

7. Wastes pass from the fetal side, to the a._____ side and b._____ pass from the maternal side to the fetal side, but there is no c._____ of blood at the placenta.

143

8. Label the following diagram of fetal circulation, using the alphabetized list of terms.
   arterial duct, oval opening, placenta, umbilical arteries, umbilical vein, venous duct

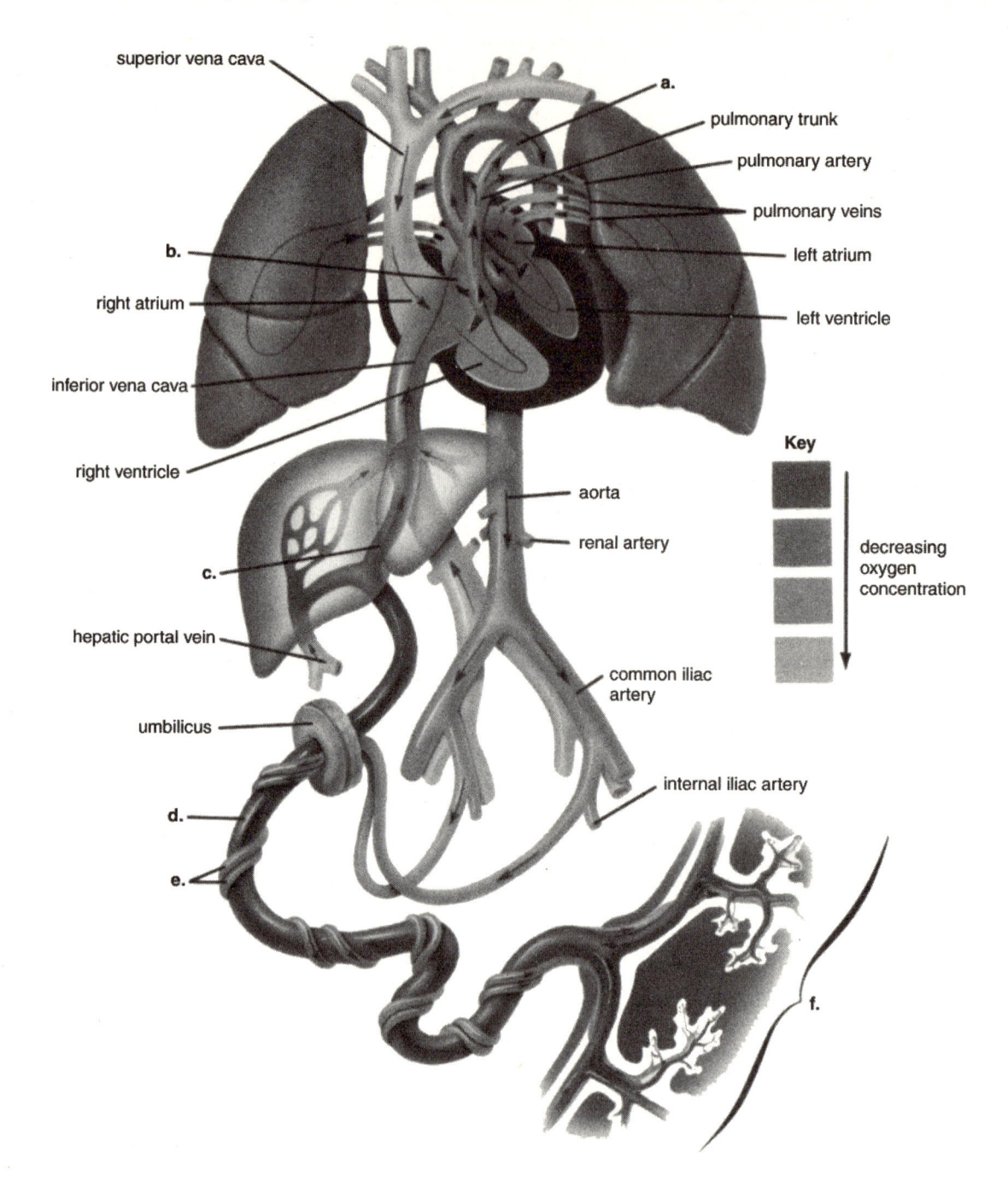

9. Trace the path of blood in the fetus from the placenta to the aorta by way of the oval opening:

   a. _____

   By way of the arterial duct: b. _____

   What vessel in the fetus carries the most oxygen? c. _____

144

10. To describe human embryonic development, complete the following table with these events:
    a. all internal organs formed; limbs and digits well formed; recognizable as human but still quite small
    b. fertilization; cell division begins
    c. limb buds begin; heart is beating; embryo has a tail
    d. implantation; embryo has tissues; first two extraembryonic membranes
    e. face taking shape; fingers and toes forming; cartilaginous skeleton forming
    f. nervous system begins; all extraembryonic membranes present; placenta well formed
    g. head enlarges; sense organs prominent

Time	Events
First week	
Second week	
Third week	
Fourth week	
Fifth week	
Sixth week	
Two months	

11. To describe human fetal development, complete the following table with these events.
    a. subcutaneous fat deposition; eyes open; testes descended
    b. possible to distinguish sex; bony skeleton is visible; head prominent; movement may be felt
    c. skin wrinkled and red, protective cheesy coating; heartbeat can be heard
    d. ready for birth

Time	Events
Third-fourth month	
Fifth-sixth month	
Seventh-eighth month	
Ninth month	

17.3 Birth (pages 376–377)

- Describe the three stages of parturition (birth).
- Describe the anatomy of the breast, and give the names and functions of the hormones that affect the breast.

12. What events are associated with the three stages of parturition?

    first stage a. _____

    second stage b. _____

    third stage c. _____

13. The incision made to prevent tearing during childbirth is called a(n) a._____. The cervix must first undergo b._____ before it dilates. True labor is marked by contractions that occur regularly every c._____ minutes.

14. A female breast contains ducts and a._____. The hormone b._____ stimulates a breast to produce milk. The hormone c._____ stimulates milk letdown.

17.4 HUMAN DEVELOPMENT AFTER BIRTH (PAGES 378–380)

• State three theories of aging, the effect of aging on the body systems, and ways to prevent a decline in body functions.

15. Three theories of aging: Some believe that aging is a._____ in origin, meaning that our b._____ inheritance causes us to age. Others maintain that c._____ processes are involved; for example, the hormonal system and the immune system decrease in efficiency as we age. Still others believe that d._____ factors influence aging more than we realize; for example, diet and exercise keep us healthy despite added years.

16. Aging affects body systems. Certain systems maintain the body; in regard to the circulatory system, a._____ disease may be associated with b._____ blood pressure, and reduced blood flow to the c._____ may result in less efficiency at filtering wastes. In regard to those systems that integrate and coordinate the body, actually few d._____ are lost from the brain, and the elderly can learn new material; loss of skeletal mass and osteoporosis can be controlled by e._____. In regard to the reproductive system, there is a reduced level of f._____ in both males and females, although males produce sperm until death. Young people should be aware that now is the time to begin the health habits that increase the life span.

DEFINITIONS WORDSEARCH

Review key terms by completing this wordsearch, using the following alphabetized list of terms:

```
D F E R G T H Y J U J Z P
C G E R O N T O L O G Y O
O U Y T R F G H A D F G A
L D S W E F R T N E R O M
O U M B I L I C U S D T N
S I O U K Y H R G S S E I
T M N B J O O P O T Y H O
R D P A R T U R I T I O N
U R E I Y T H R D C W S E
M H O E P I S I O T O M Y
V N I O Y R B M E A Q W S
```

amnion
chorion
colostrum
embryo
episiotomy
gerontology
lanugo
parturition
umbilicus
zygote

a. _____ Study of aging.
b. _____ Extraembryonic membrane that develops into placenta.
c. _____ Extraembryonic membrane that surrounds the bag of waters.
d. _____ First milk.
e. _____ Diploid cell formed after union of two gametes.

f. _____ Fetal hair forming in last trimester.
g. _____ Navel.
h. _____ Process of giving birth.
i. _____ Incision to enlarge vagina while giving birth.
j. _____ Organism during first eight weeks of development.

Chapter Test

Objective Test

Do not refer to the text when taking this test.

____ 1. During the process of fertilization
   a. one sperm only enters the egg and fuses with its nucleus.
   b. several sperm enter the egg but only one fuses with its nucleus.
   c. many sperm release acrosome enzymes so that one sperm can enter the egg.
   d. Both *a* and *c* are correct.

____ 2. The hormone that is the basis for the pregnancy test is
   a. estrogen.
   b. follicle-stimulating hormone.
   c. human chorionic gonadotropin (HCG).
   d. progesterone.

For questions 3–6, match the extraembryonic membrane with the correct description in the key.
   Key *a. placenta   b. umbilical blood vessels*
   c. watery sac   d. first site of red blood cell formation

____ 3. chorion
____ 4. amnion
____ 5. allantois
____ 6. yolk sac
____ 7. Transport occurs at the placenta by the
   a. exchange across embryonic blood vessels.
   b. mixing of maternal and embryonic blood.
____ 8. Each is true of the chorionic villi except that they
   a. are projections from an extraembryonic membrane.
   b. stay within the fetal side of the placenta.
   c. extend into the maternal portion of the placenta.
   d. initially project from the entire chorion during development.
____ 9. Which blood vessel in the fetus carries the most oxygen?
   a. pulmonary vein
   b. umbilical vein
   c. arterial duct
   d. venous duct

____ 10. The oval opening is an opening that allows fetal blood to pass from the
   a. right atrium to the left atrium.
   b. left atrium to the right atrium.
   c. right ventricle to the left ventricle.
   d. pulmonary artery into the aorta.

____ 11. The fetal circulation is modified due to the fact that
   a. the heart is not beating.
   b. the lungs are not functioning.
   c. nutrients are obtained from the yolk.
   d. All of these are correct.

____ 12. The zygote begins to undergo cleavage in the
   a. cervix.
   b. ovary.
   c. oviduct.
   d. uterus.

____ 13. Which of these is mismatched?
   a. cleavage–cell division
   b. morphogenesis–organ systems present
   c. differentiation–specialization of cells
   d. growth–increase in size

____ 14. Which is not an event normally observed by the sixth week of embryonic development?
   a. face and neck forming
   b. limb buds developing
   c. ossification begins
   d. tail regressing

____ 15. Select the incorrect association of fetal development.
   a. third month–eyes formed, lids fused
   b. fourth month–skin is bright pink and transparent
   c. fifth month–internal organs maturing
   d. sixth month–180 mm (7 in), 113 gm (4 oz)

For questions 16–18, match the stage of development in the key with the correct description.
   Key *a. embryonic development*
   b. fetal development   c. both

____ 16. heartbeat can be heard by physician
____ 17. internal organs start developing
____ 18. extraembryonic membranes formed and functioning

___19. The placenta serves to
   a. bring blood to the developing fetus.
   b. allow exchanges of substances between mother's blood and fetal blood.
   c. form the umbilical cord.
   d. Both *a* and *b* are correct.
   e. All of these are correct.

___20. Each is a normal characteristic for the seventh month during fetal development except
   a. eyes are open.
   b. fine, downy hair has largely disappeared.
   c. skin red and covered with wrinkles.
   d. testes descended into scrotal sacs.

___21. When an embryo is clearly recognizable as a human being, it is called a
   a. developed embryo.
   b. fetus.
   c. newborn.
   d. blastocyst.

___22. Which system is the first to be visually evident?
   a. nervous
   b. respiratory
   c. digestive
   d. skeletal

___23. The fontanels close after birth generally by
   a. 5–8 months.
   b. 9–13 months.
   c. 12–15 months.
   d. 16–24 months.

___24. During which stage of parturition is the baby born?
   a. first
   b. second
   c. third
   d. fourth

___25. Which hormone has no effect on the breasts?
   a. oxytocin
   b. prolactin
   c. estrogen
   d. aldosterone

THOUGHT QUESTIONS

Answer in complete sentences.

26. Why is it important for the extraembryonic membranes to develop early during human development?

27. If differentiation is not due to the parceling out of genes, what is it due to?

Test Results: _____ number correct ÷ 27 = _____ × 100 = _____%

Answer Key

Study Questions

1. a. oviduct **b.** zona pellucida **c.** acrosome **d.** fertilization **e.** vitelline **f.** implant **g.** HCG (human chorionic gonadotropin) **2. a.** cleavage **b.** morula **c.** blastocyst **d.** inner cell mass **e.** embryonic disk **f.** three **3. a.** cleavage **b.** growth **c.** morphogenesis **d.** differentiation **4. a.** chorion **b.** amnion **c.** embryo **d.** allantois **e.** yolk sac **f.** fetal portion of placenta **g.** maternal portion of placenta **5. a.** nutrient and gas exchange; becomes placenta **b.** fluid environment; cushioning, protection **c.** becomes umbilical cord for gas and nutrient exchange **d.** first site of red blood cell formation **6. a.** fetal **b.** chorion **c.** maternal **d.** uterine **7. a.** maternal **b.** nutrients **c.** mixing **8. a.** arterial duct **b.** oval opening **c.** venous duct **d.** umbilical vein **e.** umbilical arteries **f.** placenta **9. a.** umbilical vein, venous duct, vena cava, right atrium, oval opening, left atrium, left ventricle, aorta **b.** umbilical vein, venous duct, vena cava, right atrium, right ventricle, pulmonary artery, arterial duct, aorta **c.** the umbilical vein **10.** b, d, f, c, g, e, a **11.** b, c, a, d **12. a.** dilation of cervix **b.** mother pushes as baby moves down and out of birth canal **c.** afterbirth is expelled **13. a.** episiotomy **b.** effacement **c.** 15–20 **14. a.** alveoli **b.** prolactin **c.** oxytocin **15. a.** genetic **b.** genetic **c.** whole-body **d.** extrinsic **16. a.** cardiovascular **b.** high **c.** kidneys **d.** neurons **e.** exercise **f.** hormones

Definitions Wordsearch

```
                    Z
    C G E R O N T O L O G Y
    O           A       G A
    L           N       O M
    O U M B I L I C U S T N
    S           H       E I
    T           O       O O
    R   P A R T U R I T I O N
    U           I
    M   O E P I S I O T O M Y
    N       O Y R B M E
```

a. gerontology **b.** chorion **c.** amnion **d.** colostrum **e.** zygote **f.** lanugo **g.** umbilicus **h.** parturition **i.** episiotomy **j.** embryo

Chapter Test

1. d **2.** c **3.** a **4.** c **5.** b **6.** d **7.** a **8.** b **9.** b **10.** a **11.** b **12.** c **13.** b **14.** c **15.** d **16.** b **17.** a **18.** c **19.** b **20.** b **21.** b **22.** a **23.** d **24.** b **25.** d **26.** The placenta, which supplies the oxygen and nutrient needs of the embryo and fetus, is derived from the chorion, an extraembryonic membrane **27.** Although each cell in the human body has a full complement of genes, only certain of these are active in each cell. The differentiation of a cell is dependent on which genes are active.

PART 6 HUMAN GENETICS

18

CHROMOSOMAL INHERITANCE

STUDY TIPS

This chapter examines whole chromosomes and how they are passed from a parent cell to daughter cells through mitosis (pp. 391–393) and from one generation to the next through meiosis (pp. 394–399). Take a sheet of poster paper and prepare a chart contrasting the stages of mitosis with those of meiosis. Keep in mind that mitosis produces cells for growth or repair, while meiosis produces the gametes that participate in reproduction. Meiosis also has crossing-over (p. 394), a means by which new genetic combinations are possible.

On your chart, make a list of chromosomal abnormalities (pp. 387–389) that occur with some frequency in humans (e.g., Down syndrome) as a result of nondisjunction (p. 398). Include a description of the symptoms of each abnormality.

STUDY QUESTIONS

Study the text section by section as you answer the questions that follow.

18.1 CHROMOSOMAL INHERITANCE (PAGES 387–389)

- Describe a normal karyotype of a human being.
- Describe the karyotype and characteristics of Down syndrome, cri du chat syndrome, fragile X syndrome, Turner syndrome, Klinefelter syndrome, triplo-X syndrome, and Jacob syndrome.

1. a. How many chromosomes are in the human karyotype? _____
   b. How many homologous pairs are in the karyotype? _____
   c. (1) How many pairs of sex chromosomes? _____
      (2) What are the sex chromosomes for males? _____
      (3) For females? _____
   d. How many pairs of autosomes are in the karyotype? _____
   e. Which parent determines the gender of the offspring? _____
2. Complete the following table.

Name	Chromosome Abnormality	Description
Down syndrome	a.	
Fragile X syndrome	b.	
Turner syndrome	c.	
Klinefelter syndrome	d.	
Triplo-X syndrome	e.	
Jacob syndrome	f.	

150

18.2 HUMAN LIFE CYCLE (PAGE 390)

- Describe the human life cycle.

3. Complete this table to contrast mitosis with meiosis.

Cell Type	Cell Division	Description	Result
Somatic (body) cells	a.		
Sex organs	b.		

18.3 MITOSIS (PAGES 391–393)

- Draw and explain an overview of mitosis.
- Describe the cell cycle.
- Describe the stages and importance of mitosis.

4. Use the following diagram of mitosis to answer questions *a–e*.

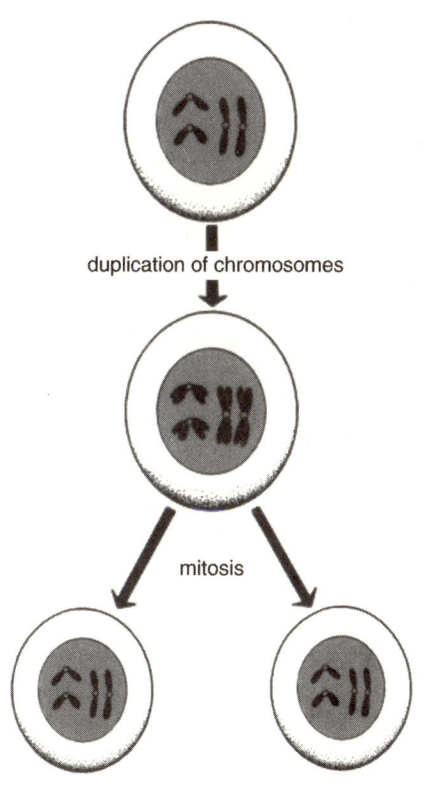

Each chromosome is composed of two identical parts called a._____. The chromatids are held together at a region called the b._____. What is the diploid number of chromosomes for the cells in the figure? c._____ Which of the cells have the diploid number? d._____
Explain why 2n → 2n is the correct way to symbolize mitosis. e._____

5. What is the cell cycle? a._____
What is the longest part of the cell cycle? b._____ What significant event occurs during this phase?
c._____

151

6. After examining the drawing of a cell in interphase (after replication has occurred), draw the stages of mitosis. Finish your drawings with daughter cells (before replication has occurred).

Mitosis

_____ prophase _____

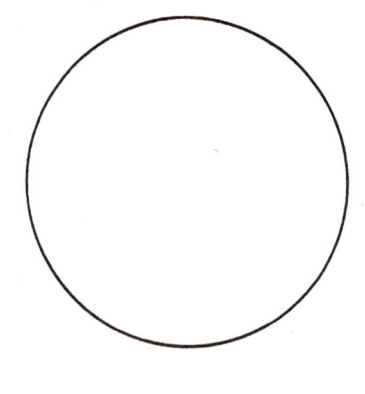

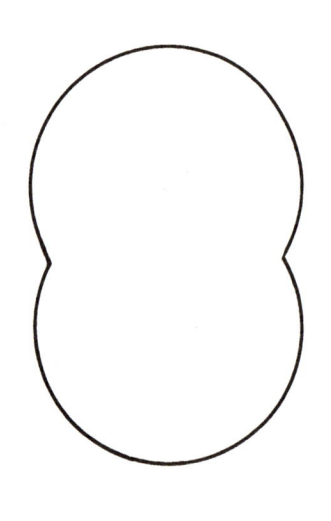

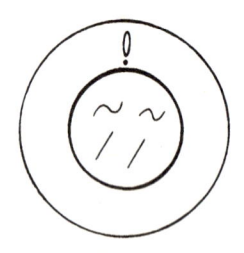

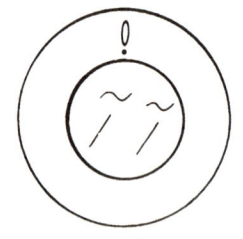

_____

_____

_____ daughter cells _____

152

18.4 Meiosis (pages 394–399)

- Draw and explain an overview of meiosis.
- Describe the stages and importance of meiosis.
- Relate chromosomal abnormalities to nondisjunction.
- Describe the process of spermatogenesis and oogenesis.

7. Use the following diagram of meiosis to answer questions *a–g*.

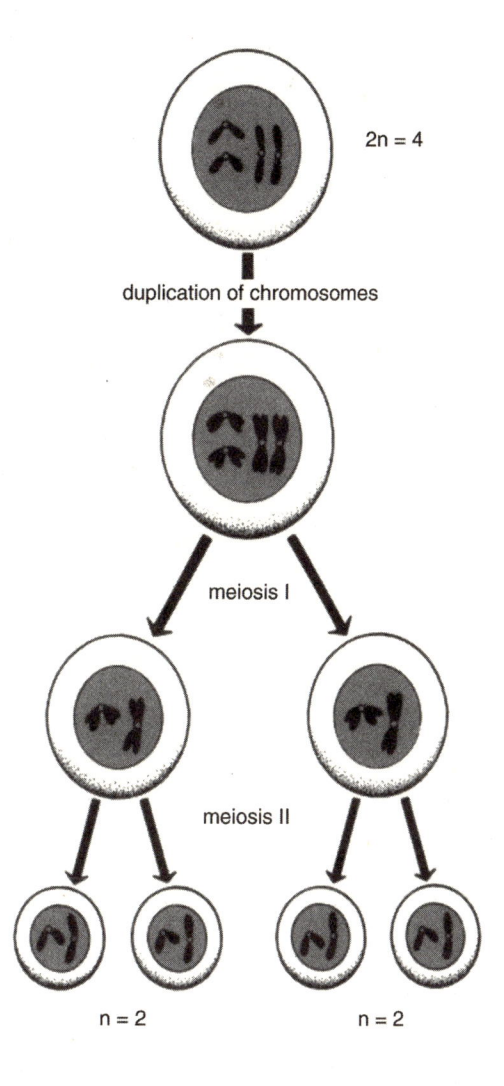

What is the diploid number of chromosomes for the cells in the figure? a._____ Does the parent cell have the diploid or the haploid number of chromosomes? b._____ Do the final daughter cells have the diploid or the haploid number of chromosomes? c._____ Explain why 2n → n is the correct way to symbolize meiosis. d._____
What structures separate during meiosis I? e._____ What structures separate during meiosis II? f._____ In humans, meiosis occurs during the production of g._____.

8. a. Draw the stages of meiosis for an animal whose diploid number is four. For meiosis I, use the diagram provided on text page 396. For meiosis II, use the diagram provided on page 397.

Meiosis I

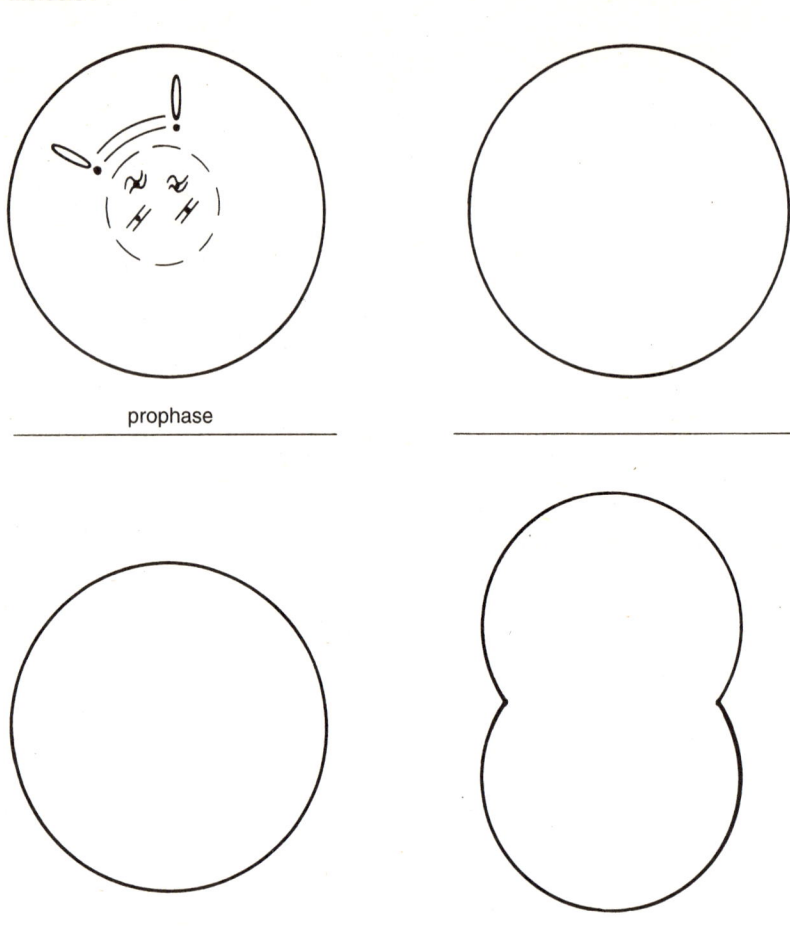

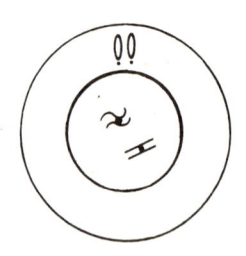

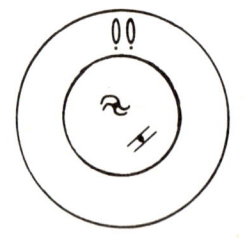

prophase

daughter cells

154

Meiosis II

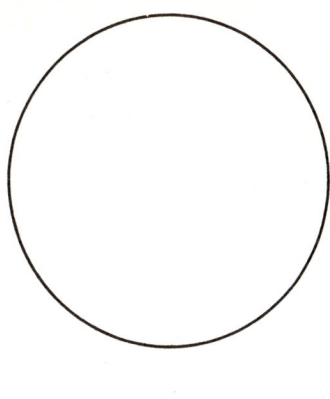

prophase
(one cell only)

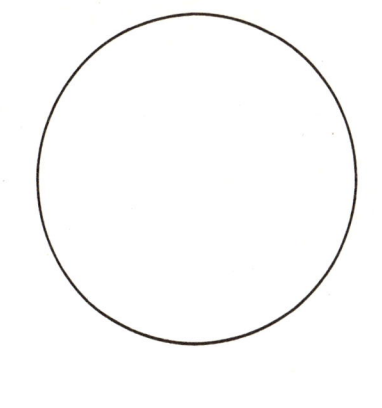

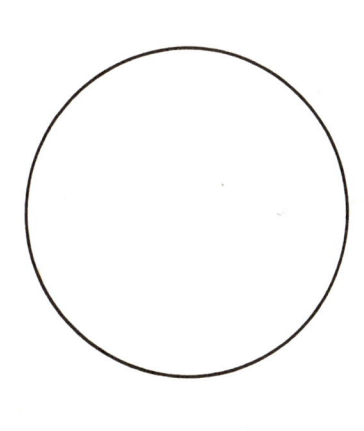

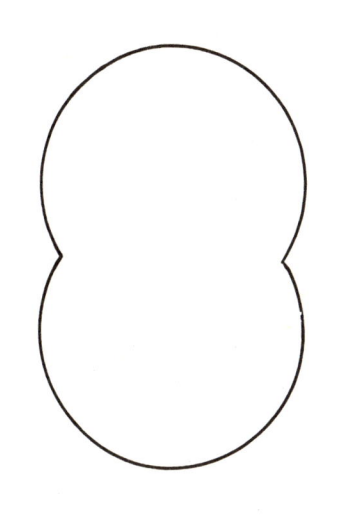

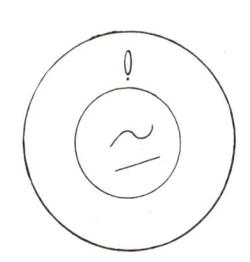

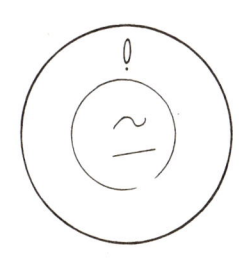

daughter cells

b. During interphase prior to meiosis I, synapsis and sometimes crossing-over occur. Using ink for one duplicated chromosome and pencil for the other, draw a tetrad before and after crossing-over has occurred.

9. Use this diagram to show how nondisjunction could bring about an abnormal egg and a Down syndrome karyotype.

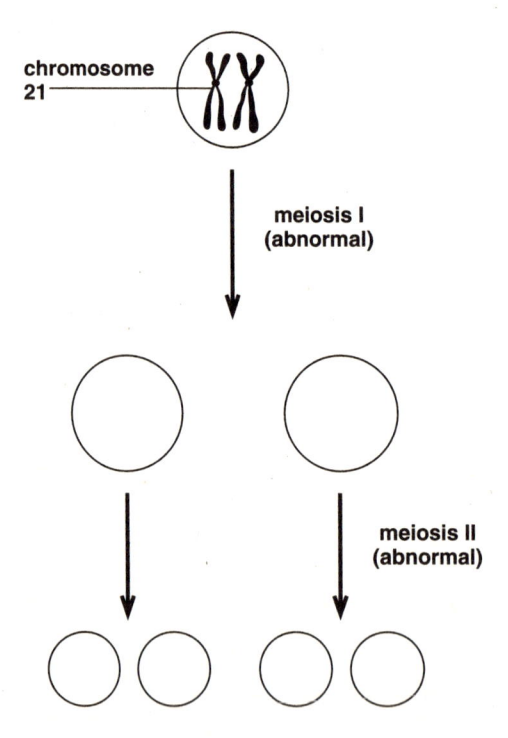

chromosome 21

meiosis I (abnormal)

meiosis II (abnormal)

10. State whether this process occurs in males, females, or both.

   production of gametes a._____

   spermatogenesis b._____

   oogenesis c._____

   polar body formation d._____

11. Fill in this table with *yes* or *no* to distinguish meiosis from mitosis.

	Meiosis	Mitosis
Completed after one division	a.	
Requires two successive divisions	b.	
During all anaphases, chromatids separate	c.	
During anaphase I, homologous chromosomes separate	d.	
Results in daughter cells with the diploid number of chromosomes	e.	
Results in daughter cells with the haploid number of chromosomes	f.	
Unique to the somatic (body) cells	g.	
Unique to the sex organs	h.	

Fill in the blanks with either *mitosis* or *meiosis* for questions 12–25.

12. _____ Occurs when a fertilized egg grows to be a baby in a mother's womb.
13. _____ Occurs when you cut your finger and it heals.
14. _____ Occurs in the testes of the male and begins in the ovary of the female.
15. _____ diploid → diploid
16. _____ diploid → haploid
17. _____ Accounts for the fact that no child looks exactly like either parent.

Refer to the drawings below to answer questions 18 and 19.

18. _____ Should be associated with the resulting daughter cell on the left in this drawing.
19. _____ Should be associated with the resulting daughter cell on the right of this drawing.
20. _____ Daughter cells have exactly the same genes as the parent cell.
21. _____ Offspring has one-half the genes of the mother or the father.
22. _____ Daughter cells have two chromosomes of each kind and two genes of each kind.
23. _____ Daughter cells have one chromosome of each kind and one gene of each kind.
24. _____ 2n → n
25. _____ 2n → 2n

Solitaire

1. Make playing cards with the following words or phrases on them:
    1. *nucleolus*
    2. *chromosomes (chromatids held by centromere)*
    3. *centrosome with centrioles*
    4. *nucleus (nuclear envelope)*
    5. *spindle fibers*
    6. *cell plate*
    7. *aster*
    8. *furrowing*
   a. From these, select those that have to do with the formation and structure of the spindle.
   b. From the cards remaining, select those that name structures that disappear during mitosis.
   c. Select the cards that have to do with cytokinesis.

2. Make cards with the following phrases on them:
    1. *chromatids separate*
    2. *distinct chromosomes within daughter nuclei*
    3. *chromosomes arranged at the metaphase plate*
    4. *chromosomes are distinct; spindle appears; nucleolus disappears and nuclear envelope fragments*
    5. *homologous chromosomes separate*
   a. Arrange the cards to describe the events of mitosis or meiosis II.
   b. Arrange the cards to describe the events of meiosis I.

3. Make three cards that are marked as follows:
    16
    8 (diploid number)
    4 (haploid number)
   a. Pick the card that tells how many chromosomes are in the parent nucleus before division.
   b. Pick the card that tells how many chromosomes each daughter nucleus will have after mitosis.
   c. Pick the card that tells how many chromosomes are in the parent nucleus during prophase of mitosis or prophase I of meiosis.
   d. Pick the card that tells how many chromatids are in the parent nucleus during prophase of mitosis or prophase I of meiosis. (For purposes of the game, assume the chromosomes are duplicated.)
   e. Pick the card that tells how many chromosomes are at *both* poles during anaphase of mitosis.
   f. Pick the card that tells how many chromosomes are at *both* poles during anaphase I of meiosis.
   g. Pick the card that tells how many chromosomes are in each daughter nucleus after meiosis I.
   h. Pick the card that tells how many chromosomes are in each daughter nucleus after meiosis II.

Definitions Wordsearch

Review key terms by completing this word search, using the following alphabetized list of terms:

```
S E S A H P R E T N I G C V F
A W A S E A D W R F T H E G Y
U E S Y N A P S I S G F N N U
T D I S D E F L G T J Y T P L
O P S L I K J U O G T F R K O
S T O G Y H U J K I H H O Y U
O R T T D I P L O I D S M B N
M P I J I J H S I S O I E M S
E G M O O G E N E S I S R R D
B V Y H F R F R D V C X E F C
S P E R M A T O G E N E S I S
```

autosome
centromere
diploid
haploid
interphase
meiosis
mitosis
oogenesis
spermatogenesis
synapsis

a. _____ Cell division resulting in cells identical to the parental cell.
b. _____ Cell division resulting in cells with half as many chromosomes as the parental cell.
c. _____ Process of egg production in females.
d. _____ Process of sperm production in males.
e. _____ Chromosome other than a sex chromosome.
f. _____ Region where sister chromatids remain attached.
g. _____ Stage of life cycle between cell divisions.
h. _____ Pairing of homologous chromosomes during prophase I.
i. _____ Half the usual number of chromosomes.
j. _____ The normal full complement of chromosomes.

Chapter Test

Objective Test

Do not refer to the text when taking this test.

___ 1. Which phrase best describes the human karyotype?
   a. 46 pairs of autosomes
   b. one pair of sex chromosomes and 23 pairs of autosomes
   c. X and Y chromosomes and 22 pairs of autosomes
   d. one pair of sex chromosomes and 22 pairs of autosomes

___ 2. Who determines the sex of the offspring?
   a. male
   b. female
   c. both male and female
   d. alternately male, then female

___ 3. Autosomes
   a. determine sex of individual.
   b. are nonsex chromosomes.
   c. refer to human chromosomes.
   d. are abnormal chromosomes.

___ 4. Which of these is an autosomal abnormality?
   a. Turner syndrome
   b. Down syndrome
   c. Klinefelter syndrome
   d. triplo-X syndrome

___ 5. XYY (Jacob syndrome) males occur, most possibly due to nondisjunction during
   a. oogenesis.
   b. spermatogenesis.
   c. fertilization.
   d. mitosis.

___ 6. Which condition is more likely to occur when the mother is over age 40?
   a. Turner syndrome
   b. triplo-X syndrome
   c. Down syndrome
   d. Klinefelter syndrome

___ 7. The *Gart* gene is
   a. on chromosome 21.
   b. associated with Down syndrome.
   c. associated with mental retardation.
   d. All of these are correct.
___ 8. Which statement is true for both genes and chromosomes?
   a. come in pairs
   b. separate during meiosis
   c. all possible combinations in gametes
   d. gametes have one-half the total number
   e. All of these are correct.
___ 9. Which of these is true of meiosis?
   a. n → n
   b. n → 2n
   c. 2n → n
   d. 2n → 2n
___ 10. DNA synthesis occurs during
   a. interphase.
   b. prophase.
   c. metaphase.
   d. All of these are correct.
___ 11. If a parent cell has 56 chromosomes, following mitosis each daughter cell will have _____ chromosomes.
   a. 14
   b. 28
   c. 56
   d. The number cannot be determined.
___ 12. The zygote
   a. is produced by fertilization.
   b. is diploid.
   c. undergoes mitosis.
   d. All of these are correct.
___ 13. Synapsis is associated with
   a. crossing-over.
   b. mitosis.
   c. meiosis II.
   d. All of these are correct.
___ 14. Normal growth and repair of the body require
   a. mitosis.
   b. meiosis.
   c. both of these are correct.
   d. Neither of these is correct.
___ 15. Which of these is involved in asexual reproduction?
   a. mitosis
   b. meiosis
   c. oogenesis
   d. spermatogenesis
___ 16. When do chromosomes move to opposite poles?
   a. prophase
   b. metaphase
   c. anaphase
   d. telophase

___ 17. The following drawing could represent a stage only in the process of
   a. replication.
   b. meiosis.
   c. mitosis.
   d. oogenesis.
   e. meiosis and oogenesis.

___ 18. By the end of meiosis I,
   a. crossing-over has occurred.
   b. chromatids of each chromosome have separated.
   c. synapsis of homologous chromosomes has occurred.
   d. each cell produced is genetically identical to the original cell.
   e. Both *a* and *c* are correct.
___ 19. During anaphase of meiosis II,
   a. homologous pairs of chromosomes separate.
   b. chromatid pairs separate.
   c. chromatids separate.
   d. duplicated chromosomes separate.
___ 20. During which stage do tetrads line up at the equator of the spindle?
   a. mitosis–metaphase
   b. meiosis–metaphase I
   c. meiosis–metaphase II
   d. Both *b* and *c* are correct.
___ 21. Meiosis
   a. keeps genetic makeup constant.
   b. is involved in asexual reproduction.
   c. assures genetic variation.
   d. prevents genetic variation.
___ 22. Which of these syndromes is/are due to nondisjunction during meiosis?
   a. Down syndrome
   b. Turner syndrome
   c. triplo-X syndrome
   d. Klinefelter syndrome
   e. All of these are correct.
___ 23. Nondisjunction is the
   a. exchange of chromosome pieces.
   b. loss of a piece of chromosome.
   c. repeating of the same piece of a chromosome.
   d. failure of chromosome pairs to separate completely.

___24. Spermatogenesis
   a. requires meiosis.
   b. produces four sperm per parent cell.
   c. occurs continuously.
   d. All of these are correct.

___25. Polar bodies are formed during
   a. meiosis.
   b. mitosis.
   c. oogenesis.
   d. spermatogenesis.
   e. Both *a* and *c* are correct.

THOUGHT QUESTIONS

Answer in complete sentences.

26. Down syndrome is due to nondisjunction of chromosomes. Explain.

27. Crossing-over increases genetic variability. Why?

Test Results: _____ number correct ÷ 27 = _____ × 100 = _____%

ANSWER KEY

STUDY QUESTIONS

1. a. 46 **b.** 23 **c.** (1) one (2) XY (3) XX **d.** 22 **e.** male parent **2. a.** extra chromosome 21; short stature, fold of upper eyelids, large fissured tongue, round head, mental retardation **b.** X chromosome is partially broken; hyperactive or autistic, jaw prominent, large head, lax joints **c.** single X; short, broad stature, ovaries are nonfunctional **d.** two or more X chromosomes and a Y; sterile, testes underdeveloped, breast development **e.** more than two X chromosomes; menstrual irregularities **f.** an X and two Y chromosomes; taller than average, persistent acne **3. a.** mitosis; 2n (diploid); 2n (diploid) growth, repair **b.** meiosis; 2n (diploid); n (haploid) gamete production **4. a.** chromatids **b.** centromere **c.** four **d.** all **e.** parent cell is 2n and daughter cells are 2n **5. a.** the cycle of events for a dividing cell **b.** interphase **c.** DNA replication **6.** see Figure 18.8, page 392, in text **7. a.** four **b.** diploid **c.** haploid **d.** parent cell is 2n and daughter cells are n **e.** homologous chromosomes **f.** chromatids **g.** gametes **8. a.** refer to text p. 396–397 **b.** see text p. 394 **9.** refer to p. 398 in the text **10. a.** both **b.** males **c.** females **d.** females **11. a.** no; yes **b.** yes; no **c.** no; yes **d.** yes; no **e.** no; yes **f.** yes; no **g.** no; yes **h.** yes; no **12.** mitosis **13.** mitosis **14.** meiosis **15.** mitosis **16.** meiosis **17.** meiosis **18.** meiosis **19.** mitosis **20.** mitosis **21.** meiosis **22.** mitosis **23.** meiosis **24.** meiosis **25.** mitosis

SOLITAIRE

1. a. 3, 5, 7 **b.** 1, 4 **c.** 6, 8 **2. a.** 4, 3, 1, 2 **b.** 4, 3, 5, 2
3. a. 8 **b.** 8 **c.** 8 **d.** 16 **e.** 16 **f.** 8 **g.** 4 **h.** 4

DEFINITIONS WORDSEARCH

```
        E S A H P R E T N I   C
A               A             E
U     S Y N A P S I S         N
T               L             T
O     S         O             R
S               O       I     O
O     T   D I P L O I D       M
M     I           S I S O I E M
E   M O O G E N E S I S R
                          E
  S P E R M A T O G E N E S I S
```

a. mitosis **b.** meiosis **c.** oogenesis **d.** spermato-genesis **e.** autosome **f.** centromere **g.** interphase **h.** synapsis **i.** haploid **j.** diploid

CHAPTER TEST

1. d **2.** a **3.** b **4.** b **5.** b **6.** c **7.** d **8.** e **9.** c **10.** a **11.** c **12.** d **13.** a **14.** a **15.** a **16.** c **17.** e **18.** e **19.** c **20.** b **21.** c **22.** e **23.** d **24.** d **25.** c **26.** Down syndrome is caused by an extra chromosome 21 in the gamete. This can occur because chromosome pairs failed to separate during meiosis I. **27.** Because of crossing-over, there is a different mix of genes on chromosomes. Therefore, the gametes are more variable than they would be if separation alone occurred.

19

GENES AND MEDICAL GENETICS

STUDY TIPS

What traits we inherit from our parents and pass on to our children is a fascinating area of study and the subject of this chapter. Master the terminology *first* (heterozygous vs. homozygous, dominant vs. recessive—see p. 404), then begin to work problems. The first step to problem solving is to write out genotypes and phenotypes (see p. 406). Then, list possible gametes *before* you try to cross them. It is a common error to try to cross two genotypes between individuals rather than crossing their gametes. Next, use a Punnett square (p. 406) to set up your cross. On some problems it is not necessary to use a Punnett square, but it is good to get into the habit of doing so. Then, work as many problems as your time allows. Learn the dominant (p. 407) and recessive disorders (p. 408), and know the examples of polygenic traits (p. 411) and multiple alleles (p. 412). Among the more interesting anomalies that occur are those carried on the sex chromosomes (pp. 414–417). Include them in your study as well.

STUDY QUESTIONS

Study the text section by section as you answer the questions that follow.

19.1 GENOTYPE AND PHENOTYPE (PAGE 404)

- Distinguish between phenotype and genotype, dominant and recessive, and homozygous and heterozygous.

1. Give the genotype of an individual who is heterozygous for earlobe attachment (E = unattached earlobe, e = attached). _____

2. What are the gametes for this person? _____

19.2 DOMINANT/RECESSIVE TRAITS (PAGES 405–409)

- Solve one-trait autosomal genetics problems using a Punnett square.
- List a variety of simple dominant and recessive traits in humans.
- Describe disorders that are inherited as dominant genes.
- Describe disorders that are inherited as recessive genes.
- Be able to interpret pedigree charts.

3. Using this key, E = unattached, e = attached, do Punnett squares for cross 1 and cross 2.

Cross 1
heterozygous × homozygous recessive

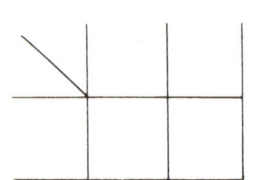

Cross 2
heterozygous × heterozygous

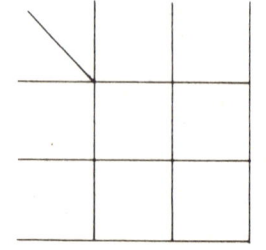

162

a. What is the phenotypic ratio for cross 1? _____

b. What are the chances of the recessive phenotype for cross 1? _____

c. What is the phenotypic ratio for cross 2? _____

d. What are the chances of the recessive phenotype for cross 2? _____

4. A man with straight hairline has children with a woman with widow's peak whose father has a straight hairline and whose mother had widow's peak. This man and woman produce a child with straight hairline. What are the genotypes of all the individuals involved?
   Key W = widow's peak   w = straight hairline

   a. man with straight hairline ____ ____

   b. woman with widow's peak ____ ____

   c. child ____ ____

   d. woman's father ____ ____

   e. woman's mother ____ ____

5. Match the disorders in the key with the phrases that follow.
   Key cystic fibrosis   Huntington disease   neurofibromatosis   phenylketonuria (PKU)   Tay-Sachs disease

   neurological impairment early in life a._____

   benign skin tumors b._____

   mental retardation prevented by special diet c._____

   neuromuscular degeneration later in life d._____

   thick mucus in lungs e._____

6. A person heterozygous for Huntington disease reproduces with a person who is perfectly normal. What are the chances of an offspring developing Huntington disease when older? _____

7. Mary is 50 years old and has Huntington disease. Her father was killed accidentally at a young age; her mother is 80 years old and is normal. What is the most likely genotype of all persons involved? Use the alleles A and a.

   Mary's mother a._____

   Mary's father b._____

   Mary c._____

8. Both parents appear to be normal, but their child has cystic fibrosis. What is the genotype of all persons involved? Use the alleles A and a.

   a. parents _____   b. child _____

For questions 9 and 10, indicate the genotype of each person in the pedigree charts. First, decide if the disorder (darkened shapes) is dominant or recessive. Use the alleles A and a in each case.

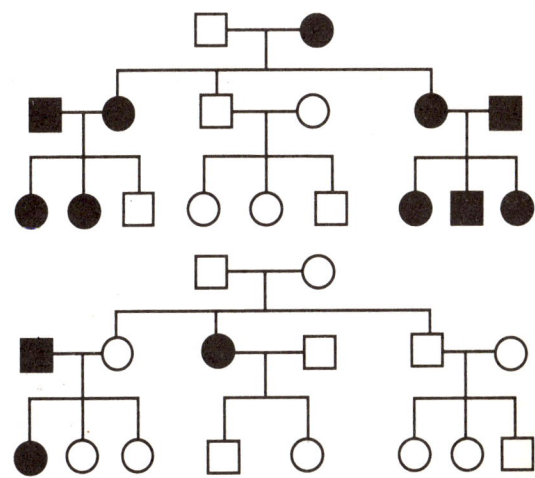

9.

10.

19.3 POLYGENIC TRAITS (PAGE 411)

• Discuss one trait inherited as a polygenic trait.

11. A woman has a genotype of *AABB* for skin color. She reproduces with a man whose genotype is *aabb*. What are the genotype and phenotype of their offspring?
    a. genotype _____   b. phenotype _____

19.4 MULTIPLE ALLELIC TRAITS (PAGE 412)

• Explain what is meant by multiple alleles, and give an example.

12. Mrs. Doe and Mrs. Roe had babies at the same hospital. Mrs. Doe took home a girl, Nancy, and Mrs. Roe received a boy, Richard. However, Mrs. Roe was certain she gave birth to a girl and brought suit against the hospital. Blood tests showed that Mr. Roe was type O, Mrs. Roe was type AB, and Mr. and Mrs. Doe were both type B. Nancy was type A, and Richard was type O. List the possible genotypes of each individual, and then indicate the correct parents for each child.
    Genotypes:

    a. Mrs. Doe _____   d. Mrs. Roe _____

    b. Mr. Doe _____   e. Mr. Roe _____

    c. Nancy _____   f. Richard _____

    g. Whose baby is Nancy? _____

    h. Whose baby is Richard? _____

19.5 INCOMPLETELY DOMINANT TRAITS (PAGES 412–413)

• Explain what is meant by incomplete dominance, and give an example.

13. Curly and straight hair, when crossed, are an example of _____ dominance because the resulting individuals have the intermediate characteristic of wavy hair.

14. A curly-haired man has children with a wavy-haired woman. What are the genotypes and phenotypes of their children?
    Key H = curly allele H' = straight allele

    Man's genotype a._____ Woman's genotype b._____
    Use this Punnett square to determine the outcome of this cross.

    Genotypes of children c._____

    Possible phenotypes of children d._____

    Could any of the children have straight hair? e._____

15. A child has sickle-cell disease. What are the genotypes of the parents, who appear to be normal? _____

19.6 SEX-LINKED TRAITS (PAGES 414–417)

- Show how sex-linked traits are generally inherited on the X chromosome.
- Discuss several sex-linked disorders.
- Distinguish between sex-linked and sex-influenced traits.

16. a. Indicate the genotype of each person in the following pedigree chart. Use alleles *A* or *a* attached to an X chromosome in each case.

    Key
    ○ normal female
    □ normal male
    ● affected female
    ■ affected male

    b. How do you know that this is a pedigree chart for an X-linked recessive trait? _____
    _____

17. Match the disorders in the key to the phrases that follow.
    Key *hemophilia   muscular dystrophy   color blindness*

    muscle weakness a. _____

    can't see reds and greens b. _____

    bleeder's disease c. _____

18. Use the Punnet square to show the expected outcome if a color-blind woman reproduces with a man who has normal vision.

    What are the chances of a color-blind daughter? a. _____

    What are the chances of a color-blind son? b. _____

19. Why are more men bald than women? _____

DEFINITIONS WORDSEARCH

Review key terms by completing this wordsearch, using the following alphabetized list of terms:

```
M E P O L Y G E N I C
U F H T H J K L O P A
L G E N O T Y P E U R
T C N F B T F D E S R
I O O X L I N K E D I
P P T I K Y H T G E E
L Y Y E D S R F G T R
E B P U N N E T T J U
F T E L E L L A P O L
```

allele
carrier
genotype
multiple
phenotype
polygenic
Punnett
X-linked

a. _____ Alternative form of a gene on a chromosome.
b. _____ Outward expression of a gene.
c. _____ Pattern of inheritance in which many genes control one trait.
d. _____ Normal individual who carries a recessive allele.
e. _____ Genetic makeup of an individual.
f. _____ Type of square used to determine genetic outcome.
g. _____ Carried on the X chromosome.
h. _____ More than two alleles for one trait are present in the population. (_____ allele)

Chapter Test

Objective Test

Do not refer to the text when taking this test.

___ 1. If 25% of the offspring of one set of parents show the recessive phenotype, the parents were probably
   a. both homozygous recessive.
   b. both homozygous dominant.
   c. both heterozygous.
   d. one homozygous dominant, one homozygous recessive.

___ 2. Alleles
   a. are alternate forms of a gene.
   b. have the same position on a pair of chromosomes.
   c. affect the same trait.
   d. All of these are correct.

___ 3. Which of these crosses could produce a blue-eyed child? (B = brown, b = blue)
   a. $BB \times bb$
   b. $Bb \times Bb$
   c. $bb \times Bb$
   d. $Bb \times BB$
   e. $Bb \times bb$

Questions 4 and 5 are concerned with this pedigree chart. The shaded individuals are affected by a disorder, whereas the unshaded individuals are not affected.

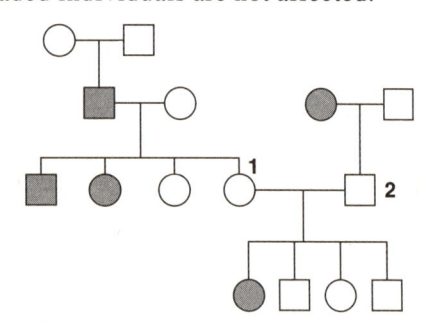

___ 4. The disorder is
   a. X-linked.
   b. dominant.
   c. recessive.
   d. not able to be determined.

___ 5. Individuals 1 and 2 are
   a. $AA \times AA$
   b. $aa \times aa$
   c. $Aa \times Aa$
   d. $Aa \times aa$
   e. $Aa \times AA$

___ 6. If a man is a carrier of Tay-Sachs disease, but a woman is homozygous normal, what are the chances of their child having Tay-Sachs?
   a. none
   b. 50%
   c. 25%
   d. 1:1

___ 7. Why is it that two normal parents have a child with PKU?
   a. PKU is a dominant inherited disorder.
   b. PKU is a recessive inherited disorder.
   c. PKU results due to an error in gamete formation.
   d. There is no known explanation.

___ 8. If only one parent is a carrier for Huntington disease, what is the chance a child will have the condition?
   a. 25%
   b. 50%
   c. 75%
   d. no chance

___ 9. Polygenic inheritance can explain
   a. a range in phenotypes among the offspring.
   b. the occurrence of degrees of dominance.
   c. the inheritance of behavioral traits.
   d. Both *a* and *c* are correct.

___ 10. Sickle-cell disease illustrates
   a. dominance.
   b. recessiveness.
   c. incomplete dominance.
   d. multiple pairing.

___ 11. Two individuals with medium-brown skin color could have children who are both darker and lighter than they are.
   a. true
   b. false

___ 12. A female with light brown skin would be able to have a child with very dark skin if she reproduces with a very dark-skinned male, or a dark child if she reproduces with a light-skinned male.
   a. true
   b. false

___ 13. Which children could not have parents both with type A blood?
   a. type A
   b. type O
   c. type AB
   d. type B
   e. Both *c* and *d* are correct.

___ 14. Which children could not have a parent with type AB blood?
   a. type A
   b. type B
   c. type AB
   d. type O

___ 15. Inheritance by multiple alleles is illustrated by the inheritance of
   a. skin color.
   b. blood type.
   c. sickle-cell disease.
   d. Both *b* and *c* are correct.

___16. When may complications arise regarding a pregnancy?
   a. Rh⁻ woman and Rh⁺ man
   b. Rh⁺ woman and Rh⁻ man
   c. Rh⁺ woman and Rh⁺ man
   d. Rh⁻ woman and Rh⁻ man

Questions 17 and 18 pertain to this pedigree chart.

___17. The allele for this disorder is
   a. dominant.
   b. recessive.
   c. X-linked.
   d. None of these is correct.

___18. The genotype of the starred individual is
   a. Aa
   b. aa
   c. $X^A X^a$
   d. $X^A Y^a$

___19. A woman who is not color blind but has an allele for color blindness reproduces with a man who has normal color vision. What is the chance they will have a color-blind daughter?
   a. 50%
   b. 25%
   c. 100%
   d. no chance

___20. A color-blind woman reproduces with a man who has normal color vision. Their sons will
   a. be like the father because the trait is X-linked.
   b. be like the mother because the trait is X-linked.
   c. all have normal color vision.

Match questions 21–24 to the following key.
   Key a. hemophilia   b. sickle-cell disease
   c. Tay-Sachs disease   d. Huntington disease

___21. X-linked disorder
___22. autosomal recessive
___23. abnormal red blood cells
___24. symptoms appear later in life
___25. A girl is color blind (sex-linked recessive trait).
   a. She received a color-blind allele from her mother.
   b. She received a color-blind allele from her father.
   c. All her sons will be color blind.
   d. Her father is color blind.
   e. All of the above are correct.

THOUGHT QUESTIONS

Answer in complete sentences.

26. To test whether an animal has a homozygous dominant genotype or a heterozygous genotype, it is customary to mate them to the homozygous recessive animal rather than a heterozygote. Why?

27. Males who have a sex-linked genetic disorder most likely inherited the abnormal allele from their mothers. Why?

Test Results: _____ number correct ÷ 27 = _____ × 100 = _____%

Answer Key

Study Questions

1. *Ee* **2.** *E* and *e* **3. a.** 1:1 **b.** 50% **c.** 3:1 **d.** 25% **4. a.** *ww* **b.** *Ww* **c.** *ww* **d.** *ww* **e.** *Ww* **5. a.** Tay-Sachs **b.** neurofibromatosis **c.** PKU **d.** Huntington disease **e.** cystic fibrosis **6.** 50% **7. a.** *aa* **b.** *Aa* **c.** *Aa* **8. a.** *Aa* **b.** *aa* **9. top:** *aa, Aa* **middle:** *Aa, Aa; aa; aa; A? A?* **bottom:** *A? A? aa; aa, aa, aa; A?; A?, A?* **10. top:** *Aa, Aa* **middle:** *aa, Aa; aa, A?; A?, A?* **bottom:** *aa, Aa, Aa; Aa, Aa; A?, A?, A?* **11. a.** *AaBb* **b.** medium brown **12. a.** *BB* or *BO* **b.** *BB* or *BO* **c.** *AA* or *AO* **d.** *AB* **e.** *OO* **f.** *OO* **g.** Nancy belongs to the Roes. **h.** Richard belongs to the Does. **13.** incomplete **14. a.** *HH* **b.** *HH'* **c.** *HH* or *HH'* **d.** curly or wavy hair **e.** no, because one parent must pass on an *H* **15.** both are heterozygous for the trait **16. a. top:** X^aY, X^AX^A **middle:** X^AX^A, X^AY, X^AX^A, X^AY, X^AX^a, X^AY **bottom:** X^AY, X^AY, X^AY, X^AX^A, X^AX^A, X^aY **b.** Only men have the trait, and it skips from grandfather to grandson. **17. a.** muscular dystrophy **b.** color blindness **c.** hemophilia **18. a.** 0% **b.** 100% **19.** Men need only one allele for baldness, while women need two to be bald.

Definitions Wordsearch

```
M  P O L Y G E N I C
U  H             A
L  G E N O T Y P E R
T  N             R
I  O X L I N K E D I
P  T             E
L  Y             R
E  P U N N E T T
E L E L L A
```

a. allele **b.** phenotype **c.** polygenic **d.** carrier **e.** genotype **f.** Punnett **g.** X-linked **h.** multiple

Chapter Test

1. c **2.** d **3.** b, c, e **4.** c **5.** c **6.** a **7.** b **8.** b **9.** d **10.** c **11.** a **12.** b **13.** e **14.** d **15.** b **16.** a **17.** c **18.** c **19.** d **20.** b **21.** a **22.** c **23.** b **24.** d **25.** e **26.** If a heterozygote is mated to a heterozygote, there is a 25% chance for any offspring to be recessive. If a heterozygote is mated to a homozygous recessive, there is a 50% chance for any offspring to be recessive. It is only when the animal is mated to a homozygous recessive that its genotype can be determined. **27.** Males inherit only a Y sex chromosome from their fathers. The Y chromosome is blank for most sex-linked traits, and therefore, the abnormal allele must be on the X chromosome inherited from the mother.

20

DNA and Biotechnology

STUDY TIPS

In this chapter, you will learn about the exciting field of biotechnology. However, you must first understand the basic structure and function of the nucleic acids that form the foundation for this field.

Compare the structure of DNA (p. 423–424) with that of RNA (p. 425). Remember that in biology, structure relates to function. Learn how the structure of DNA allows it to be the genetic material for the cell, and how RNA can be the messenger between the DNA and the cytoplasm of the cell.

When you study the section on gene expression (pp. 426–431), draw a step-by-step diagram illustrating each event from transcription to translation to protein synthesis. You may need to review it several times to have a thorough understanding of gene expression. Understand how the cell controls gene expression (p. 431).

Study the diagrams indicating how recombinant DNA is manufactured. Produce your own diagram if need be. Study the polymerase chain reaction (p. 433), and understand how it helps mass-produce copies of DNA. Be ready to give several examples of how this reaction is employed in biology and other fields.

Make note of a number of genetically engineered organisms produced thus far (pp. 434–437), and how gene therapy is employed in humans (p. 437).

STUDY QUESTIONS

Study the text section by section as you answer the questions that follow.

20.1 DNA AND RNA STRUCTURE AND FUNCTION (PAGES 422–425)

- Describe the structure of DNA.
- Explain what is meant by complementary base pairing.
- Understand how DNA replicates.
- Describe the structure and function of RNA.
- List the three types of RNA found in cells.
- Discuss the similarities and differences between DNA and RNA.

DNA Structure and Replication (pages 423–424)

1. Place a *T* beside the statements that are true concerning DNA structure and *F* by those that are false. Explain or rewrite where indicated.
   _____ a. DNA is a nucleic acid found in the nucleus.
   _____ b. DNA is a polynucleotide. Explain. _____
   _____
   _____ c. Each nucleotide in DNA is composed of the sugar ribose, a phosphate, and a base. Rewrite: _____
   _____
   _____ d. DNA is a double-stranded helix.

169

2. Examine the "ladder" structure of DNA and answer the questions that follow.

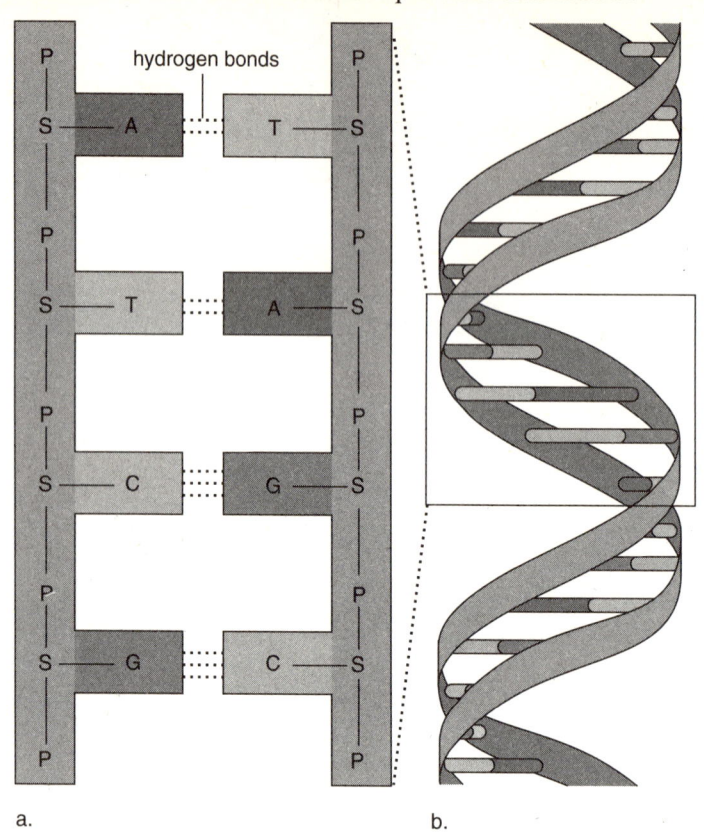

a.

b.

Which types of molecules make up the sides of the ladder? a. _____ and b. _____
Which types of molecules make up the rungs of the ladder? c. _____
Explain complementary base pairing. d. _____
What type of bond joins the bases? e. _____
What do you do to the ladder structure to make a double helix? f. _____

3. When DNA replicates, the hydrogen bonds break and the two strands are unzipped. Assume that replication has occurred and complete this diagram by drawing in complementary strands.

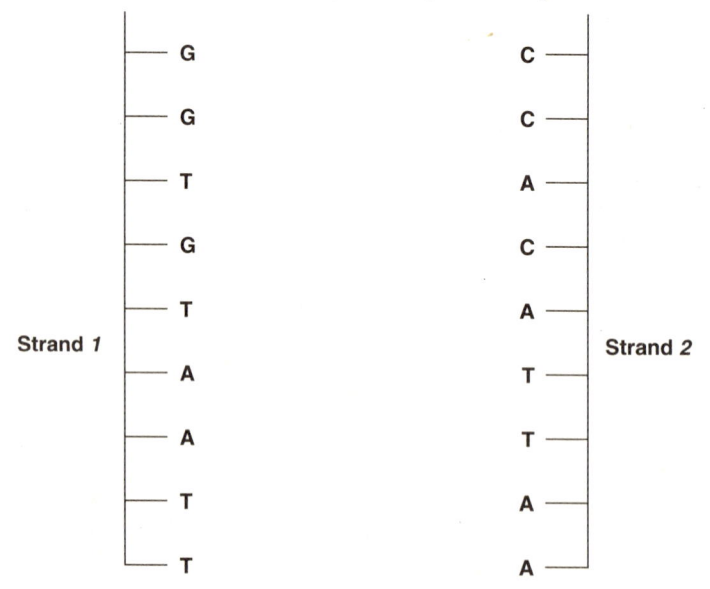

Strand 1: G, G, T, G, T, A, A, T, T

Strand 2: C, C, A, C, A, T, T, A, A

In your drawing, does the double strand on the left resemble the double strand on the right? a. _____
Explain. _____

170

Why is replication called semiconservative? b. _____

_____

If a mistake occurs during replication and it is not corrected, a c. _____ has taken place.

The Structure and Function of RNA (Page 425)

4. Since genes (DNA) reside in the a. _____ of the cell and polypeptide synthesis occurs in the b. _____, they must have a go-between. The most likely molecule to fill this role is c. _____.

5. Indicate whether the following statements about differences between DNA and RNA are true (T) or false (F):
   _____ a. DNA is double stranded; RNA is single stranded.
   _____ b. DNA is a polymer; RNA is a building block of that polymer.
   _____ c. DNA occurs in three forms; RNA occurs in only one.
   _____ d. The sugar of DNA is ribose, which is absent in RNA.
   _____ e. Uracil, in RNA, replaces the base thymine, found in DNA.

9. Complete this table to describe the function of the various types of RNA involved in protein synthesis.

RNA	Function
messenger RNA (mRNA)	a. _____
ribosomal RNA (rRNA)	b. _____
transfer RNA (tRNA)	c. _____

20.2 GENE EXPRESSION (PAGES 426–431)

- Describe the structure and function of proteins.
- Explain what is meant by DNA's triplet code.
- Describe transcription and processing of messenger RNA (mRNA).
- Describe translation of mRNA codons.
- Understand how the cell controls gene expression.

Structure and Function of Proteins (Page 426)

7. The following diagram represents portions of two different proteins.

Val — His — Tyr — His — Ala — Val — Leu
 1   2   3   4   5   6   7
His — Val — Leu — Ala — His — Tyr — Ala

Each circle in the diagram represents a(n) a. _____. Each and every protein has its own particular sequence of b. _____.

Would the amino acid at number 3 necessarily be the same in both proteins? c. _____ At any number would the amino acids necessarily be the same? d. _____ How is one protein different from another? e. _____

Proteins have structural and functional roles. Name a protein whose role is primarily structural. f. _____ How do proteins function in metabolic pathways? g. _____

The DNA Code (Page 426)

8. Whereas DNA has a triplet code—that is, every three a. _____ stand for a(n) b. _____. —mRNA contains c. _____, each of which is made up of three bases that also stand for an amino acid.

171

Transcription (Pages 426–427)

9. In the nucleus, DNA serves as a template for the production of messenger RNA (mRNA). Let us assume that strand 2 in the diagram for question 2 will serve as the template. What will be the codons that mRNA brings to the ribosomes?
   a. _____

   Before mRNA leaves the nucleus, what happens to it? b. _____
   _____

Translation (Pages 428–429)

10. Label the following diagram to describe translation.

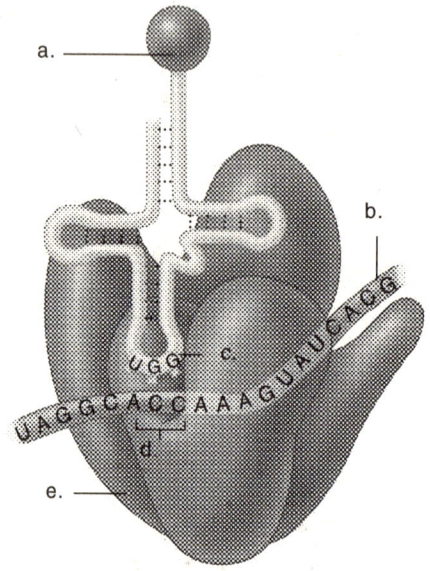

    a. _____
    b. _____
    c. _____
    d. _____
    e. _____

11. Two types of RNA are seen in the diagram for question 10. Ribosomal RNA (rRNA) plus proteins make up the ribosomes. Each ribosome is composed of a a. _____ subunit and a b. _____ subunit. Transfer RNA is the second type of RNA in the diagram. At one end an c. _____ attaches, and at the other end there is an d. _____, which is complementary to a codon in mRNA.

12. The three steps in protein synthesis are a. _____, when the ribosomal subunits and rRNA b. _____; and c. _____, when a polypeptide is d. _____; and e. _____, when the last tRNA, the mRNA, and the ribosome f. _____.

13. During elongation, the sequence of a. _____ in mRNA dictates the order of b. _____ in the polypeptide. For example, if the sequence of bases in mRNA is UUU UUA AUU GUC CCA, the sequence of amino acids in the polypeptide according to the figure in question 10 will be c. _____. Because several ribosomes called a d. _____ can move along one mRNA molecule, (one/many) e. _____ polypeptides of the same type can be synthesized at a time.

The Control of Gene Expression (Page 431)

14. Complete the following table:

Levels of Control of Gene Activity	Affects the Activity of

172

15. What are transcription factors? _____
_____

20.3 BIOTECHNOLOGY (PAGES 432–439)

- List and describe the means by which genes are cloned.
- Describe the polymerase chain reaction.
- Give examples of biotechnology products produced by bacteria, plants, and animals.
- List and describe the means by which animals are cloned.
- State the purposes of the Human Genome Project.
- Describe two possible methods for gene therapy in humans.

16. In the diagram, write the numbers of the following descriptions in the appropriate blanks:
    1 Cloning occurs when host cell reproduces.
    2 Host cell takes up recombined plasmid.
    3 DNA ligase seals human gene and plasmid.
    4 Restriction enzyme cleaves DNA.

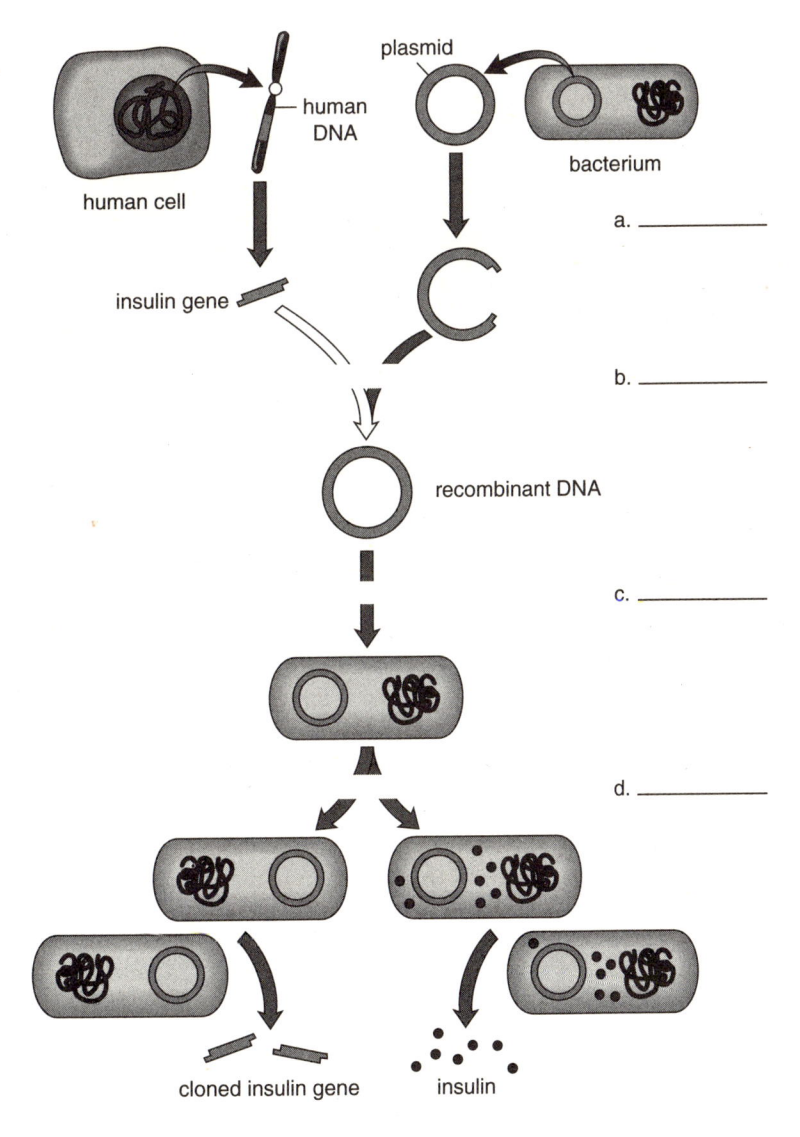

a. _____

b. _____

c. _____

d. _____

17. What is meant by the expression that restriction enzymes produce "sticky ends"? _____
_____

173

18. Change the following false statements to true statements:

    a. Only plasmids are used as vectors during genetic engineering experiments. Rewrite: _____

    b. Recombinant DNA contains two types of bacterial DNA recombined together. Rewrite: _____

    c. Genetic engineering usually means that an organism receives genes from a member of its own species. Rewrite: _____

    d. Gene cloning occurs when a gene produces many copies of various genes. Rewrite: _____

19. Use the following diagram to help you complete this paragraph.

    The clones on the petri dish are from a genomic library. What is a genomic library? a._____
    If these clones are from a mammalian cell, the genes are accompanied by b._____ regions that are necessary for the expression of mammalian genes. If these clones have been made using reverse transcriptase, then the cDNA contains the c._____ but not the introns. In the diagram, a radioactive probe is being added to the petri dish. What is a probe? d._____

20. Explain the polymerase chain reaction by telling what *polymerase* refers to: a._____; and what chain reaction means: b._____. At the beginning of the reaction, very little DNA may be available, but at the end of the reaction c._____ copies of a segment of DNA are available.

21. In DNA fingerprinting, a._____ enzymes digest the two samples to be compared. b._____ separates the fragments, and their different lengths are compared. If the pattern is similar, the samples are from c._____.

22. Complete the following table on transgenic organisms:

Type of Organism	Engineered for What Purpose

23. The advantage of using bacteria to produce a product is that a._____.
    The advantage of using plants to produce a product is that b._____.
    The advantage of using farm animals to produce a product is that c._____.

24. Which of these is true of xenotrasplantation?
    _____ a. Xenotransplantation uses humans as a source of organs for transplants.
    _____ b. Xenotransplantation uses other species, such as the pig, as a source of organs for transplants.
    _____ c. Pigs can be genetically altered to prevent the rejection of their organs by humans.
    _____ d. Other species can pass new and different viruses to humans.

25. Put these statements in the proper sequence to describe the making of a transgenic female goat that will produce a medicine needed by humans in its milk.
    _____ a. Development within a host animal
    _____ b. Remove egg from donor animal.
    _____ c. Isolate a human gene.
    _____ d. Microinject the human gene into the egg of the donor animal.
    _____ e. Transgenic goat is born.

26. Put these statements in the proper sequence to describe the cloning of the transgenic goat produced in question 25.
    _____ a. Birth of cloned transgenic goats
    _____ b. Remove nuclei from adult cells of transgenic goat.
    _____ c. Collect the milk which contains the medicine of interest.
    _____ d. Development within host goats
    _____ e. Microinject 2n nuclei into the donor eggs.
    _____ f. Remove eggs form donor animal.

27. The Human Genome Project has two goals, which are (1) to make a ᵃ·_____ map of the human chromosomes and (2) to make a ᵇ·_____ map of the human chromosome.

28. Which one of these is not a step in sequencing the base pairs of DNA?
    _____ a. Cut the DNA into fragments.
    _____ b. Use a computer program to put the fragments in the correct order.
    _____ c. Use a DNA sequencer to order the base pairs.
    _____ d. Put the DNA into bacteria for cloning the fragments.

29. After eliminating the incorrect step in the preceding question, put the others in the correct order. _____

30. State two benefits for human beings if we knew the sequence of the bases in the human genome.
    a. _____
    b. _____

31. Change these false statements to true statements.
    a. Ex vivo methods of gene therapy require that the therapeutic gene be placed in the body either directly or by using a viral vector. Rewrite: _____

    b. A common ex vivo method is to microinject normal genes into bone marrow stem cells removed from the patient. Then the stem cells are returned to the patient. Rewrite: _____

    c. Gene therapy is currently restricted to curing genetic diseases and is not used to treat illnesses like cystic fibrosis or cardiovascular diseases. Rewrite: _____

175

Gene Expression Maze

Can you find your way through the maze to a polypeptide by identifying each of the components depicted?

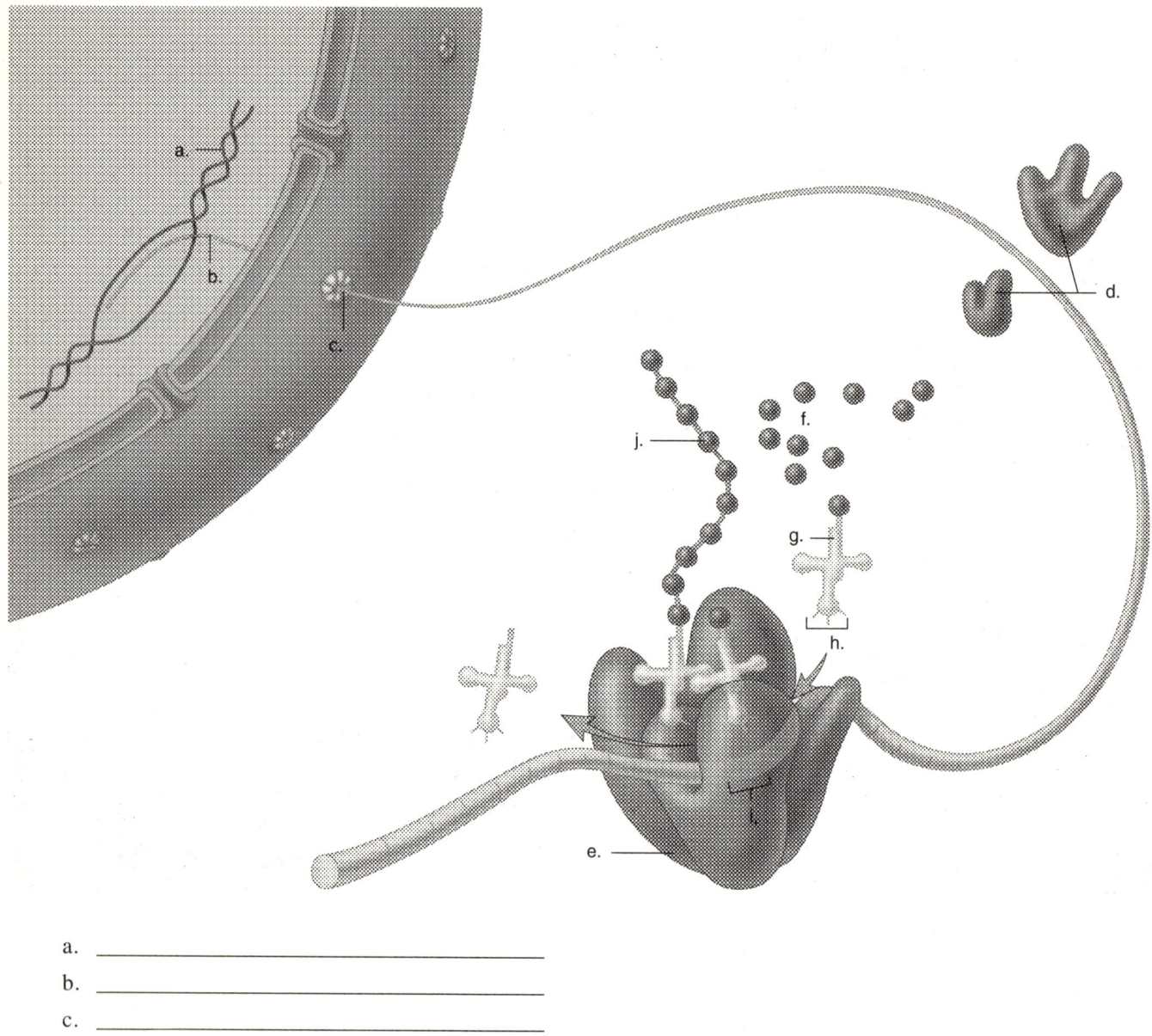

a. _____
b. _____
c. _____
d. _____
e. _____
f. _____
g. _____
h. _____
i. _____
j. _____

If you identified all correctly, you have found your way out.

DEFINITIONS WORDSEARCH

Review key terms by completing this wordsearch, using the following alphabetized list of terms:

```
G T R F Y H J D P L O I J
B I O T E C H N O L O G Y
A U Y E H T R A E C D E D
N R T M E W S L A O S N I
T P L P I J Y I R D W E M
I G Y L H F D G C O X Z S
C V B A C X Z A Z N V D A
O T Y T L O P S H Y G F L
D T Y E F R Q E A Z M L P
O T R A N S F E R R N A K
N A R E P L I C A T I O N
```

anticodon
biotechnology
codon
DNA ligase
gene
plasmid
replication
template
transfer RNA

a. _____ Unit of heredity passed on to offspring.

b. _____ Enzyme used to join DNA from two sources.

c. _____ RNA molecule that carries the amino acid to the ribosome.

d. _____ Circular fragment of DNA from bacterial cell.

e. _____ Making an exact copy, as in the duplication of DNA.

f. _____ Triplet of bases in mRNA.

g. _____ Complementary triplet of bases in tRNA.

h. _____ Use of a natural biological system to produce a commercial product.

i. _____ One strand of DNA that serves as a pattern during duplication.

CHAPTER TEST

OBJECTIVE TEST

Do not refer to the text when taking this test.

____ 1. Which is *not* true of DNA?
   a. polymer of nucleotides
   b. occurs in nucleus
   c. has a varied sequence of just three bases
   d. contains deoxyribose

____ 2. DNA is the hereditary material because it
   a. stores genetic information.
   b. replicates.
   c. controls cells' activities.
   d. All of these are correct.

____ 3. Which of the following pairs is *not* a valid comparison between DNA and RNA?

	DNA	RNA
a.	double helix	single stranded
b.	replicates	duplicates
c.	deoxyribose	ribose
d.	thymine	uracil

____ 4. A person with a metabolic defect, such as the inability to make a particular protein, inherited a
   a. faulty gene.
   b. faulty code.
   c. mutation.
   d. All of these are correct.

___ 5. If a DNA molecule has this sequence of base pairs, what will be the sequence following replication?

```
├─ A ........ T ─┤
├─ T ........ A ─┤
├─ C ........ G ─┤
├─ G ........ C ─┤
```

   a. exactly the same as this
   b. exactly opposite of this
   c. wherever a *T* appears after replication there would be a *U*
   d. like this, only each DNA molecule would be single stranded

___ 6. If mutation occurred, then
   a. the code would change.
   b. some particular codon or codons would change.
   c. some particular anticodon or anticodons would change.
   d. All of these are correct.
   e. Both *a* and *b* are correct.

___ 7. Transcription
   a. involves DNA.
   b. involves tRNA.
   c. occurs at the ribosome.
   d. All of these are correct.

___ 8. A mutation has occurred when the sequence of
   a. bases in DNA changes.
   b. bases in mRNA changes.
   c. amino acids in protein changes.
   d. All of these are correct.

___ 9. Which pair is mismatched?
   a. DNA–code
   b. mRNA–codon
   c. tRNA–anticodon
   d. rRNA–recodon

___ 10. Put phrases 1–6 in the correct order to describe gene expression.
   (1) mRNA is produced in the nucleus.   a. 1,2,3,4,5,6
   (2) Ribosomes move along mRNA.   b. 2,4,6,1,3,5
   (3) DNA has a code.   c. 3,1,6,2,5,4
   (4) Polypeptide results.   d. 3,5,1,6,2,4
   (5) tRNA brings amino acids to ribosomes.
   (6) mRNA moves to ribosomes.

___ 11. Protein synthesis takes place in the
   a. nucleus, where DNA codes for amino acids.
   b. cytoplasm, where the ribosomes code for amino acids.
   c. nucleus, where the tRNAs code for amino acids.
   d. cytoplasm, where mRNA codons pair with tRNA anticodons.

___ 12. How does the anticodon differ from the codon?
   a. The anticodon contains thymine, but the codon contains uracil.
   b. The anticodon attaches to ribosomes, but the codon attaches to amino acids.
   c. The anticodon is a sequence of three bases complementary to the bases of a codon.
   d. The anticodon stands for a particular amino acid, but the codon codes for nucleotide bases.
   e. All of these are correct.

___ 13. What two types of molecules are involved when the codon pairs with its anticodon?
   a. mRNA and DNA
   b. mRNA and tRNA
   c. DNA and tRNA
   d. rRNA and DNA

___ 14. Which of these is happening when translation takes place?
   a. mRNA is still in the nucleus.
   b. tRNAs are bringing amino acids to the ribosomes.
   c. rRNAs expose their anticodons.
   d. DNA is being replicated.
   e. All of these are correct.

___ 15. Which two events are in proper sequence?
   a. translation–replication
   b. translation–transpiration
   c. transcription–translation
   d. translation–transcription

___ 16. Which of these is a true statement concerning translation?
   a. Each polypeptide is synthesized one amino acid at a time.
   b. The amino acids are joined by RNA polymerase at the same time.
   c. Each ribosome is responsible for adding a single amino acid to each polypeptide.
   d. Since the code is degenerate, the same type of polypeptide often contains a different sequence of amino acids.
   e. All of these are correct.

___ 17. The problem with using pigs as a source of human organs is
   a. it's not possible to genetically alter pig cells to avoid rejection.
   b. their organs are too small in comparison to human organs.
   c. pig organs might carry viruses that would be new to humans.
   d. transplant patients would rather die than receive organs from a pig.

___ 18. Plasmids
   a. are found in bacteria.
   b. are rings of DNA.
   c. may be used to make recombinant DNA.
   d. All of these are correct.

___19. A biotechnology product could not be a(n)
   a. enzyme.
   b. vaccine.
   c. hormone.
   d. fat molecule.

___20. To clone a transgenic animal, you have to have
   a. an egg donor.
   b. special food to feed them.
   c. a way to protect human beings from coming in contact with transgenic animals.
   d. a transgenic animal.
   e. both *a* and *d*.

___21. The polymerase chain reaction
   a. produces many copies of different segments of DNA.
   b. produces many copies of the same segment of DNA.
   c. requires denaturing double-stranded DNA into single strands by heating.
   d. Both *b* and *c* are correct.

___22. Which of these is *not* a step to prepare recombinant DNA?
   a. remove plasmid from bacterial cell
   b. use restriction enzyme to acquire foreign gene and cut open vector
   c. use ligase to seal foreign gene into vector
   d. use a virus to carry recombinant DNA into a plasmid

___23. Which of these is a benefit to having insulin produced by biotechnology?
   a. It can be mass-produced.
   b. It is nonallergenic.
   c. It is less expensive.
   d. All of these are correct.

___24. Biotechnology vaccines are
   a. pathogens treated to be nonvirulent.
   b. proteins produced by a pathogenic gene.
   c. only enzymes taken from a pathogen.
   d. Both *a* and *b* are correct.

___25. The Human Genome Project
   a. will probably take about 50 more years more to complete.
   b. has no usefulness to humans.
   c. will result in knowing the base pair sequences of all the DNA in a human cell.
   d. must be carried out by government employees because the results are top secret.

THOUGHT QUESTIONS

Answer in complete sentences.

26. It is said that DNA stores information. What kind of information does it store?

27. Why would you expect a transgenic animal to pass the newly acquired gene of interest on to its offspring?

Test Results: _____ number correct ÷ 27 = _____ × 100 = _____ %

Answer Key

Study Questions

1. a. T **b.** T Many nucleotides are joined in DNA. **c.** F ... composed of the sugar deoxyribose **d.** T   **2. a.** phosphate **b.** sugar **c.** nitrogen-containing bases **d.** A is paired with T and G is paired with C and vice versa. **e.** hydrogen **f.** twist it   **3. a.** yes. Because of complementary pairing of the bases they have to be similar **b.** Each DNA molecule is made up of an old (parental) strand and a new (daughter) strand. **c.** mutation   **4. a.** nucleus **b.** cytoplasm **c.** RNA   **5. a.** T **b.** F **c.** F **d.** F **e.** T   **6. a.** takes a message from DNA in the nucleus to the ribosomes in the cytoplasm **b.** is found in ribosomes, where proteins are synthesized **c.** transfers amino acids to the ribosomes   **7. a.** amino acid **b.** amino acids **c.** no **d.** no **e.** sequence of amino acids **f.** muscle proteins actin and myosin **g.** as enzymes   **8. a.** bases **b.** amino acid **c.** codons   **9. a.** GGU, GUA, AUU **b.** It is processed.   **10. a.** amino acid **b.** mRNA **c.** anticodon **d.** codon **e.** ribosome   **11. a.** large **b.** small **c.** amino acid **d.** anticodon   **12. a.** initiation **b.** associate **c.** elongation **d.** lengthened **e.** termination **f.** dissociate   **13. a.** bases **b.** amino acids **c.** phenylalanine, leucine, isoleucine, valine, proline **d.** polyribosome **e.** many   **14.**

Levels of Control of Gene Activity	Affects the Activity of
transcriptional	DNA
posttranscriptional	mRNA during formation and processing
translational	mRNA life span during protein synthesis
posttranslational	protein

15. factors that must bind to DNA before transcription can begin   **16. a.** 4 **b.** 3 **c.** 2 **d.** 1   **17.** Cleavage results in unpaired bases.   **18. a.** Both plasmids and viruses are used . . . **b.** contains DNA from two different sources **c.** . . . from a member of a different species **d.** . . . many copies of the same gene   **19. a.** a collection of clones that carry all the genes of an organism **b.** regulatory **c.** exons **d.** single-stranded DNA or mRNA that hybridizes (pairs) with a gene of interest   **20. a.** DNA polymerase, the enzyme involved in DNA replication **b.** the reaction occurs over and over again **c.** many   **21. a.** restriction **b.** gel electrophoresis **c.** from the same individual

22.

Type of Organism	Engineered for What Purpose
bacteria	to protect plants, for bioremediation, to produce chemicals, and to mine metals
plants	to resist insects, pesticides and herbicides, and to make products
animals	to have improved qualities and to make products

23. a. They will take up plasmids. **b.** They will grow from single cells (protoplasts). **c.** The product is easily obtainable in milk.   **24.** b, c, d   **25.** c, b, d, a, e   **26.** b, f, e, d, a, c   **27. a.** genetic **b.** base sequence   **28.** d   **29.** a, c, b   **30. a.** We could acquire normal genes for gene therapy. **b.** We could determine the genetic illnesses of individuals.   **31. a.** In vivo **b.** . . . use a viral vector to carry normal genes. . . **c.** . . . is not restricted to curing genetic disease and is used to treat illnesses. . .

Gene Expression Maze

a. DNA **b.** mRNA **c.** nuclear pore **d.** ribosomal subunits **e.** ribosome **f.** amino acids **g.** tRNA **h.** anticodon **i.** codon **j.** peptide chain

Definitions Wordsearch

```
              D
B I O T E C H N O L O G Y
A     E     A   C   E D
N     M     L   O   N I
T     P     I   D   E M
I     L     G   O     S
C     A     A   N     A
O     T     S         L
D     E     E         P
O T R A N S F E R R N A
N   R E P L I C A T I O N
```

a. gene **b.** DNA ligase **c.** transfer RNA **d.** plasmid **e.** replication **f.** codon **g.** anticodon **h.** biotechnology **i.** template

Chapter Test

1. c **2.** d **3.** b **4.** d **5.** a **6.** e **7.** a **8.** d **9.** d **10.** c **11.** d **12.** c **13.** b **14.** b **15.** c **16.** a **17.** b **18.** d **19.** d **20.** d **21.** d **22.** d **23.** d **24.** b **25.** c **26.** The sequence of bases in DNA dictates the proper sequence of amino acids in a protein. This is the information stored in DNA.   **27.** The gene is now a part of the animal's chromosomes and is passed on with the chromosomes.

21

CANCER

STUDY TIPS

A number of fairly specific treatments are available for each of the various forms of cancer. In general, however, surgery (p. 453) is used when the tumor is localized and removing it will not cause it to spread to surrounding tissues. Surgery is also done when a tumor is disrupting the function of a particular organ. Radiation (p. 453) is a follow-up procedure to kill any remaining cancer cells after surgical removal. It can also be used to treat or shrink certain forms of cancer. Because of the possibility of cell damage from radiation, it is used only on localized areas. Cancer chemotherapy (p. 453), on the other hand, is used when it is likely that cancer has spread or to try and prevent its spread to other areas of the body. The term chemotherapy can also be applied to any form of medication when it is used to treat any ailment, not just cancer.

When studying this chapter, the most difficult portion involves understanding how cell division is controlled by protein factors within the cell. It is when something goes wrong with the genes that produce these factors that cancerous cells are formed. Carefully outline this portion of the chapter (pp. 446–448) using a flow diagram (i.e., proto-oncogenes → (code for) growth factor proteins or growth factor receptors; environmental factors → (cause) proto-oncogenes → (to become) oncogenes; and so forth).

Lastly, be aware that personal habits largely determine whether or not we develop cancer. Prevention (p. 449), routine screening (p. 451), and healthy life-style choices (p. 449) can either lessen the risk of ever developing cancer or greatly reduce its severity if it does develop.

STUDY QUESTIONS

Study the text section by section as you answer the questions that follow.

21.1 CANCER CELLS (PAGE 444)

- State and explain five characteristics of cancer cells.
- Explain how the patient's prognosis is determined by the stage of cancer, and list the three stages of progression.
- Tell how cancers are classified.

1. Complete this table, which compares normal cells with cancer cells.

Characteristics of Normal Cells Versus Cancer Cells

Characteristic	Normal Cells	Cancer Cells
Differentiation	Yes	No
Nuclei	Normal	a.
Tumor formation	Controlled	Uncontrolled
Contact inhibition	Yes	No
Growth factors	Required	b.
Angiogenesis	c.	Yes
d.	No	Yes

181

2. When determining prognosis, the physician will examine the patient for the involvement of surrounding tissues to see whether or not the cancer has spread beyond the initial tumor site. Place the following statements in the proper order of more to less favorable prognosis.
   _____ a. lymph node involvement
   _____ b. tumor invasion of surrounding tissues
   _____ c. metastatic tumors in distant areas of the body

21.2 ORIGIN OF CANCER (PAGES 446–448)

- Describe four classes of genes associated with the development of cancer.
- Describe the regulatory network that stretches from the plasma membrane to and including the nucleus.
- Describe the role of apoptosis in the development of cancer.

3. Explain why a mutation in these genes can lead to cancer.
   a. genes for repair enzymes _____
   b. genes for proteins that regulate the structure of chromatin _____
   c. proto-oncogenes and tumor-suppressor genes _____

4. Label the following diagram, using the alphabetized list of terms.
   growth factor, growth inhibitory factor, inhibitory pathway, proto-oncogene and tumor-suppressor gene, receptor protein, signaling proteins, stimulatory pathway, tumor-suppressor gene

5. In the stimulatory pathway, the reception of a ⁿ·_____ causes signaling proteins to turn on a(n) ᵇ·_____ in the nucleus. In the inhibitory pathway, reception of a(n) ᶜ·_____ causes signaling proteins to turn on a(n) ᵈ·_____ in the nucleus. When a proto-oncogene becomes a(n) ᵉ·_____ or when a tumor-suppressor gene undergoes a mutation, then uncontrolled growth threatens. The oncogenes most frequently involved in human cancers belong to the ᶠ·_____ gene family. The tumor-suppressor gene called ᵍ·_____ is frequently mutated in human cancers. This gene when functioning normally can stimulate ʰ·_____, programmed cell death, which is a safeguard to prevent the development of tumors.

21.3 CAUSES AND PREVENTION OF CANCER (PAGES 448–450)

- Discuss two factors that contribute to the development of cancer.
- List several types of carcinogens.
- Identify and discuss seven ways (including dietary guidelines) to prevent cancer from developing.

6. What are two factors that contribute to the development of cancer? ᵃ·_____
   ᵇ·_____

7. What is a carcinogen? ᵃ·_____
   Name three types of carcinogens. ᵇ·_____
   _____
   What habit exposes many people to organic chemicals? ᶜ·_____
   What medical procedure exposes many people to radiation? ᵈ·_____
   Name a virus that is known to cause cancer of the cervix. ᵉ·_____

8. Give two examples of types of cancer that have a hereditary basis. _____
   _____

9. Name three behaviors that are associated with the development of cancer. ᵃ·_____
   _____
   Name a behavior that helps prevent cancer. ᵇ·_____
   What types of foods should be included in the diet? ᶜ·_____
   _____
   What types of foods should be kept to a minimum? ᵈ·_____
   _____
   What vitamins are especially helpful in preventing cancer? ᵉ·_____
   Why? ᶠ·_____

21.4 DIAGNOSIS AND TREATMENT (PAGES 451–456)

- List and explain routine screening tests and tumor marker tests for cancer.
- List and describe ways to confirm the diagnosis of cancer.
- List and describe the standard methods of cancer therapy.
- List and describe future methods of cancer treatment.

10. You would expect these screening test to detect which type of cancer?
    Pap smear ᵃ·_____
    mammography ᵇ·_____
    sigmoidoscopy ᶜ·_____
    stool blood test ᵈ·_____

11. Tumor marker tests are a._____ tests that depend upon the fact that the tumor b._____ provoke a(n) c._____ response in the body.

12. Which of these imaging techniques requires the administration of a radioactive isotope? _____
    a. CAT scan b. MRI c. ultrasound

Treatment of Cancer

13. Three standard methods of treatment are a._____, b._____, and c._____.

    Explain the rationale behind the use of radiation. d._____

    Explain the rationale behind the use of chemotherapy e._____

    Explain the rationale behind bone marrow transplants f._____

14. Identify these treatments as *radiation, chemotherapy,* or *bone marrow transplants.*
    _____ a. Patient's stem cells are harvested and stored before treatment begins.
    _____ b. Proton beams are aimed at the tumor like a rifle bullet hitting the bull's eye of a target.
    _____ c. Taxol is given to patients with ovarian cancers.

15. Identify each of these therapies as immunotherapy (*I*), gene therapy (*GT*), chemoprevention (*C*), or complementary therapies (*CT*)
    _____ a. An inhibitor of the enzyme known as COX-2 can prevent metastasis.
    _____ b. Monoclonal antibodies can be linked to radioactive isotopes.
    _____ c. Acupuncture
    _____ d. Use of the genetically engineered adenovirus kills cancer cells when it is injected into the body.
    _____ e. Antiangiogenic drugs confine and reduce tumors by breaking up the network of new capillaries in the vicinity of a tumor.

DEFINITIONS WORDSEARCH

Review key terms by completing this wordsearch, using the following alphabetized list of terms:

```
P A P S M A H I F T E R T I P
Q W O R R A M M H Y U E R D C
V C A N C E R M U R O M U T S
T H E S A S A U J K U L F E N
G H U J R F E N E G O C N O S
E T H D C B Y O R T F V I K L
R T D S I M E T A S T A S I S
A H N A N T O H F A T D O G S
C H E M O T H E R A P Y R A T
D A G N G K O R B R B A K C B
N R A O E T C A R C I N O M A
C A T G N U J P C O O K I T U
R S U T U N L Y M P H O M A P
H A M P P T H A K N G V N G I
```

cancer
carcinogen
carcinoma
chemotherapy
immunotherapy
lymphoma
metastasis
mutagen
oncogene
tumor

a. _____ Environmental agent that can produce mutations.
b. _____ Gene that helps turn a normal cell into a cancer cell.
c. _____ Growth of cells derived from one mutated cell.
d. _____ Malignant tumor that metastasizes.
e. _____ Environmental agent that contributes to the start of cancer.
f. _____ Spread of cancer.
g. _____ Cancer of the lymphoid organs.
h. _____ Use of drugs to kill off cancer cells.
i. _____ Cancer arising in epithelial tissue.
j. _____ Use of therapy derived from immune system.

Chapter Test

Objective Test

Do not refer to the text when taking this test.

____ 1. Which of these is *not* a known carcinogen?
   a. nitrates
   b. radiation
   c. water
   d. tobacco smoke

____ 2. Choose the one statement that is true.
   a. Only one-hundredth of cancer deaths can be attributed to smoking.
   b. When smoking is combined with alcohol, cancer risk increases.
   c. Eating smoked meats is good for your health.
   d. Sunbathing produces healthy, tanned skin.

____ 3. Which of these is mismatched?
   a. sunlight–skin cancer
   b. high-fat diet–breast cancer
   c. sexual activity–breast cancer
   d. smoking–lung cancer

____ 4. Factors that lead to carcinogenesis
   a. include hereditary, radiation, and carcinogens.
   b. are mostly unknown.
   c. include unsanitary conditions first and foremost.
   d. are highly correlated with drug use.

____ 5. Which of these might explain why vitamins A and C help prevent cancer?
   a. They kill bacteria in the gut.
   b. They are antioxidants.
   c. They cause the growth of cells.
   d. They kill cells.

____ 6. A healthy, cancer-preventing diet should *not* include much/many
   a. vegetables.
   b. fruits.
   c. fiber.
   d. fats.

____ 7. Chemoprevention therapies include
   a. the use of monoclonal antibodies to zero in the plasma membrane receptors of cancer cells.
   b. the use of viruses to carry tumor-suppressor genes into cancer cells.
   c. the drug COX-2 to prevent metastasis and antiangiogenic drugs to reverse angiogenesis.
   d. the use of acupuncture, biofeedback, and exotic foods to cure cancer.

____ 8. Which of these is *not* a characteristic of cancer cells?
   a. specialization of structure and function
   b. metastasis
   c. uncontrolled growth
   d. nondifferentiation

____ 9. Metastasis is
   a. involved in the initiation of cancer.
   b. a localized tumor.
   c. cured by the proper diet.
   d. the spread of cancerous cells.

____ 10. The formation of new blood vessels, as in a cancerous growth, is
   a. angiogenesis.
   b. spermatogenesis.
   c. oogenesis.
   d. metastasis.

____ 11. Which of these is mismatched?
   a. carcinoma–leukemia
   b. carcinoma–colon cancer
   c. sarcoma–bone cancer
   d. leukemia–malignant lymphocytes

____ 12. Oncogenes are
   a. viral genes.
   b. tumor-suppressor genes.
   c. always inherited.
   d. involved in the growth cycle of cells.

____ 13. Proto-oncogenes
   a. can become oncogenes by mutation.
   b. help control the growth of cells.
   c. can be carried by viruses.
   d. All of these are correct.

____14. Cancer-causing genes are
  a. often oncogenes.
  b. always involved in the growth of cells.
  c. often mutated tumor-suppressor genes.
  d. All of these are correct.

____15. Signaling proteins
  a. are the cause of cancer antigens.
  b. are always receptors.
  c. are a part of a regulatory pathway.
  d. make cells contract.

____16. Which of these is a tumor marker test?
  a. monthly breast exam to find tumors
  b. getting a mammogram for breast cancer
  c. getting a CAT scan to find tumors
  d. doing a blood test for CEA to detect colon cancer

____17. Genital warts are associated with cancer of the
  a. penis.
  b. breast.
  c. vagina.
  d. cervix.

____18. If cancer is present,
  a. the stimulatory pathway that stretches from the plasma membrane to the nucleus is active and promoting cell division.
  b. the inhibitory pathway is inactive.
  c. oncogenes and mutated tumor-suppressor genes are probably in the nucleus.
  d. apoptosis is not occurring
  e. All of these are true.

____19. Telomerase is an enzyme that
  a. is active in cancer cells.
  b. restores the telomeres in normal cells.
  c. is active in normal cells.
  d. tears down the telomeres in cancer cells.
  e. Both *a* and *d* are true.

____20. Which of these is mismatched?
  a. mammography–breast cancer
  b. Pap smear–lung cancer
  c. blood test– prostate cancer
  d. heredity–breast cancer

____21. Chemotherapy following surgery is for the purpose of
  a. supplying the body with the chemicals it needs.
  b. catching stray cells that have or might metastasize.
  c. preventing resistance from occurring.
  d. curing leukemia only.

____22. Gene therapy for cancer would
  a. use a virus as a vector.
  b. require bone marrow transplants.
  c. shut down oncogenes and supply tumor-suppressor cells.
  d. Both *a* and *c* are correct.

____23. Signaling proteins
  a. are active after a cell receives a growth factor or a growth inhibitory factor.
  b. turn on proto-oncogenes and tumor-suppressor genes.
  c. can be stimulatory or inhibitory.
  d. All of these are true.

____24. Tumor marker tests are always
  a. reliable.
  b. blood tests.
  c. available for any cancers.
  d. All of these are correct.

____25. Which type of cancer is highly curable?
  a. Hodgkin's disease
  b. advanced ovarian cancer
  c. childhood leukemia
  d. Both *a* and *c* are correct.

THOUGHT QUESTIONS

Answer in complete sentences.
26. Explain the manner in which cancer can be an inheritable disorder.

27. Explain the manner in which cancer can be a failure of the immune system.

Test Results: _____ number correct ÷ 27 = _____ × 100 = _____ %

Answer Key

Study Questions

1. a. Abnormal **b.** Not required **c.** No **d.** Metastasis
2. b, a, c **3. a.** Repair enzymes ordinarily prevent mutations during DNA replication. **b.** These proteins ordinarily cause chromatin to be available for transcription only when appropriate. **c.** This enzyme rebuilds the telomeres at the ends of chromosomes; ordinarily the telomeres shortened, preventing ongoing cell division. **d.** Proto-oncogenes and tumor-suppressor genes ordinarily regulate cell division. **4.** see Figure 21.3, in text
5. a. growth factor **b.** proto-oncogene **c.** growth-inhibitory factor **d.** tumor-suppressor gene **e.** oncogene **f.** *ras* **g.** *p53* **h.** apoptosis **6. a.** carcinogens **b.** heredity
7. a. an environmental agent that can contribute to the development of cancer. **b.** certain organic chemicals, radiation, pollutants, dietary fat, viruses **c.** smoking **d.** X rays **e.** papillomavirus **8.** some breast cancers, retinoblastoma
9. a. smoking, sunbathing, exposure to radiation or chemicals **b.** proper diet **c.** fiber, fruits, vegetables **d.** fat-laden foods, salt-cured, smoked, or nitrate-cured foods **e.** vitamins A and C **f.** They are antioxidants. **10. a.** cervical cancer **b.** breast cancer **c.** colon cancer **d.** colon cancer **11. a.** blood **b.** antigens **c.** antiody **12.** b. MRI
13. a. radiation **b.** chemotherapy **c.** bone marrow transplants **d.** Radiation is mutagenic, and dividing cancer cells are more susceptible to its effects. **e.** Chemotherapy is a way to catch cancer cells that have spread throughout the body. **f.** To allow the use of high doses of chemotherapeutic drugs, bone marrow stem cells are harvested and stored before therapy begins. **14. a.** bone marrow transplant **b.** radiation **c.** chemotherapy **15. a.** C **b.** I **c.** CT **d.** GT **e.** C

Definitions Wordsearch

```
                I
                M
    C A N C E R M   R O M U T
        A       U
        R   E N E G O C N O
        C       O
        I   M E T A S T A S I S
    N   N       H
    C H E M O T H E R A P Y
        G   G   R
        A   E   C A R C I N O M A
        T   N   P
        U       L Y M P H O M A
        M
```

a. mutagen **b.** oncogene **c.** tumor **d.** cancer **e.** carcinogen **f.** metastasis **g.** lymphoma **h.** chemotherapy **i.** carcinoma **j.** immunotherapy

Chapter Test

1. c **2.** b **3.** c **4.** a **5.** b **6.** d **7.** c **8.** a **9.** d **10.** a **11.** a **12.** d **13.** d **14.** d **15.** c **16.** d **17.** d **18.** e **19.** a **20.** b **21.** b **22.** d **23.** d **24.** b **25.** d **26.** Cancer is inheritable when parents pass on proto-oncogenes and/or oncogenes to their children. **27.** Cancer may be a failure of the immune system when T cells fail to recognize cancer cells as nonself cells and therefore do not destroy them.

Part 7 Human Evolution and Ecology

22

Evolution

Study Tips

The evidence for evolution falls into a comparative number of categories. There is the fossil record (p. 462), comparative biochemical evidence (p. 464), comparative anatomical evidence (p. 463), and biogeographical evidence (p. 464). Choose an interesting example from the text for each of these categories to help you remember how it fits in as evidence for evolution.

The evolution of humans is a fascinating topic to study. How humans came to use tools and which group exhibited the first signs of a religion are still somewhat under speculation. Make a chronological listing of the species of hominids leading up to modern humans (pp. 469–471). Beside each, list the special traits for the group (e.g., tool user, walked upright). Then, make note of how they were similar to modern humans and how they differed.

Students often wonder how scientists distinguish between ape fossils and human fossils. The scientists who specialize in the study of hominid fossils (archaeologists) are experts in identifying such traits as pelvic shape, dentition, cranial capacity, spinal curvature, and so forth, that allow them to easily distinguish between ape fossils and those that lead to humans. Humans and apes at one point shared a common ancestor, but that was millions of years ago. (It is often mistakenly thought that archaeologists argue that humans evolved from apes.)

Study Questions

Study the text section by section as you answer the questions that follow.

22.1 Evidence for Evolution (pages 462–464)

- Give examples of the evidence for the evolutionary process.
- Understand the use of the term *theory* as a great unifying principle of biology.

1. Match each description with the type of evidence in the key.
    Key biochemical   embryology   geography   history of life   homologous structures

    a. _____ All chordates have gill arches during development.
    b. _____ Human hemoglobin is like that of a chimpanzee.
    c. _____ A bat wing and a human forearm have the same bones.
    d. _____ Fossils of bacteria are older than fossils of mammals.
    e. _____ Golden jackals in Africa are adapted similarly to coyotes in North America.

2. Our knowledge of the history of life is based mostly on the a._____ record. Modern humans appear in the record about b._____ years ago. Almost all creatures use the same basic biochemical molecules, such as c._____. Anatomical structures that are developed in one group of organisms but reduced and nonfunctional in another are called d._____ structures.

3. When the term *theory* in used in conjunction with the idea of evolution, what does theory indicate? _____
_____

22.2 THE EVOLUTIONARY PROCESS (PAGE 465)

- Order and explain five statements that outline the process of natural selection as developed by Darwin.

4. Put these statements in the proper order, by rearranging the letters, to describe Darwin's theory of natural selection.
   _____ a. The end result of organic evolution is many different species, each adapted to specific environments.
   _____ b. Many more individuals are produced each generation than can survive and reproduce.
   _____ c. Gradually, over long periods of time, a population can become well adapted to a particular environment.
   _____ d. There are inheritable variations among the members of a population.
   _____ e. Individuals with adaptive characteristics are more likely to be selected to reproduce by the environment.

5. Are all members of a species exactly the same? a._____ What does it mean to say that they vary?
   b._____

6. The members of a species compete for resources. Why is competition necessary? _____
_____

7. The members of which species are more adapted to living in a forest?
   _____ a. ground plants that are able to grow in the shade
   _____ b. ground plants that require full sunlight

8. The members of which species are better adapted to living in a mountain village?
   _____ a. those who get dizzy when atmospheric oxygen is low
   _____ b. those who thrive even in low atmospheric oxygen

9. Complete the following statements:
   Darwin emphasized that better-adapted individuals a._____. Modern evolutionists emphasize that better-adapted individuals are able to b._____.

22.3 ORGANIC EVOLUTION (PAGES 466–468)

- Give a sequence of events by which the first cell may have evolved and life thereafter diversified.
- Name the classification categories in sequence and be able to recognize a binomial name.

10. Match these phrases with *a–d* in the drawing that follows.
    *simple organic molecules form   oxygen in atmosphere and aerobic cellular respiration evolves
    macromolecules lead to protocell   primitive atmosphere contains gases*

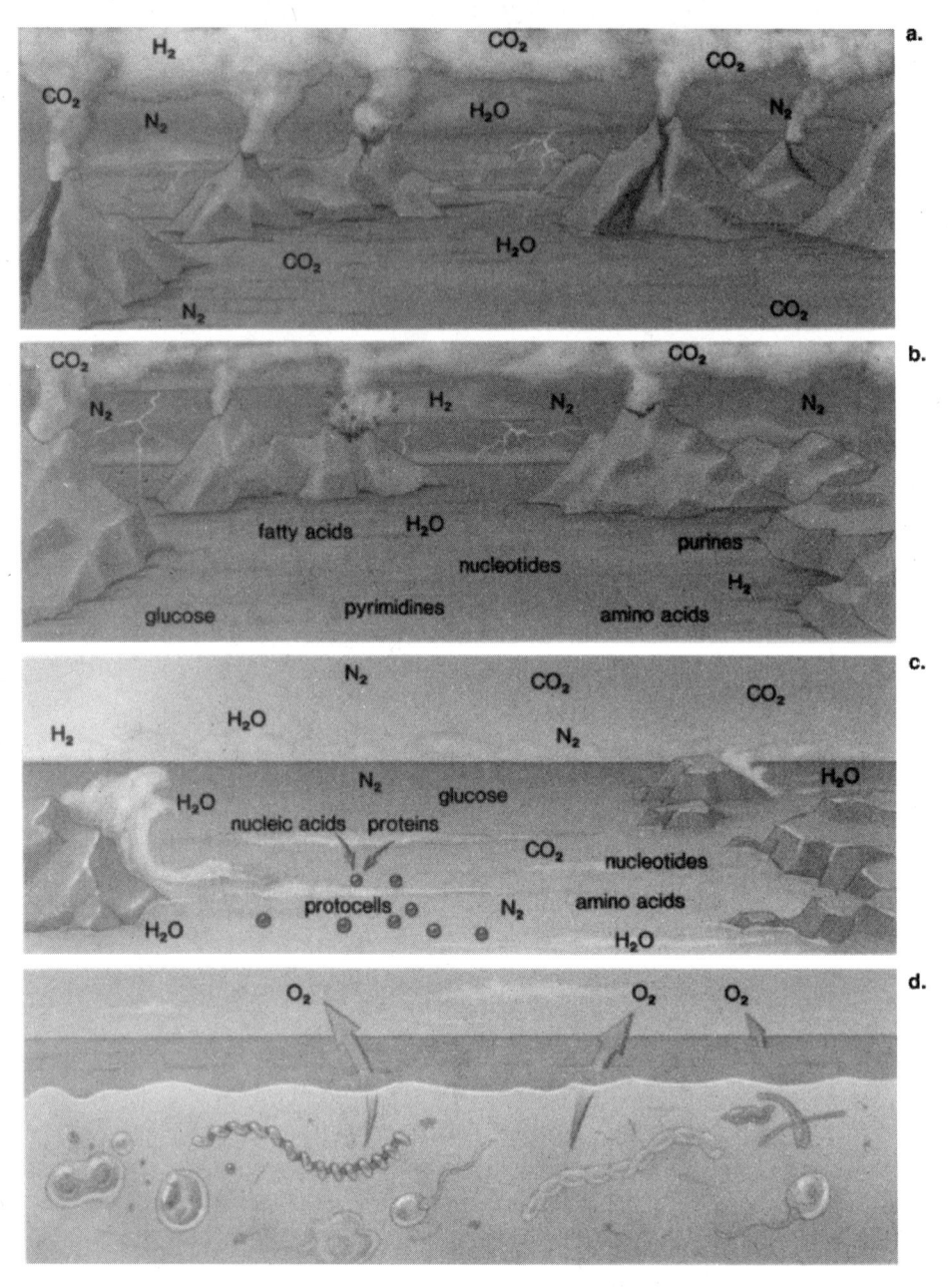

a.

b.

c.

d.

11. The early atmosphere lacked free a._____, but there was b._____ vapor, which condensed and became the oceans. The gases of the early atmosphere formed c._____ when they were exposed to outside d._____, such as heat from volcanoes and electric discharges in lightning. The oceans were now a thick e._____. The simple organic molecules combined to form f._____, and these led to the formation of a protocell. When the protocell could reproduce, it became a true cell, which must have been a(n) g._____, living on organic molecules in the ocean and breaking them down anaerobically.

12. Taxonomy. Make up a jingle that will help you to remember the order of the taxonomic categories.
    a._____
    _____

Which two categories are used in a binomial name? b._____

_____

Does taxonomy always stay the same? Explain. c._____

_____

13. Place these designations next to the correct category in order to classify humans: *sapiens,* Mammalia, Hominidae, Chordata, Animalia, Primates, *Homo.*

    kingdom a. _____

    phylum b. _____

    class c. _____

    order d. _____

    family e. _____

    genus f. _____

    species g. _____

22.4 MODERN HUMANS EVOLVE (PAGES 469–472)

- Tell what characteristics distinguish primates from other mammals.
- Trace the evolution of humans from the hominids to Cro-Magnon.
- Explain how human races represent phenotypes all belonging to the same species.

14. Associate the primate characteristics in the key with the human features listed.

    Key opposable thumb   expanded cerebrum   living in trees   long childhood

    tool use a._____ and b._____

    walking erect c._____

    intelligence d._____

    learning after birth e._____

15. Name types of animals in each category.

    mammals a._____

    primates b._____

    hominids c._____

16. Arrange these hominids in the order they evolved: *Homo habilis,* Neanderthal, *Homo erectus,* australopithecines, Cro-Magnon. Associate these characteristics with the earliest possible tool use, knowledge of fire, walking erect, modern appearance, *Homo sapiens.*

Name	Characteristic
a. _____	_____
b. _____	_____
c. _____	_____
d. _____	_____
e. _____	_____

17. The Cro-Magnon were accomplished hunters. Some researchers believe that they caused the extinction of *Homo sapiens neanderthalensis* and many uniquely large mammals. State another hypothesis to explain the disappearance of *Homo sapiens neanderthalensis.* _____

_____

_____

18. Why do we know now that human races represent phenotypes rather than genetically distinct races as was once thought?
    a. _____
    _____
    _____

    To what are these phenotypes adapted? b. _____
    _____
    _____

DEFINITIONS CROSSWORD

Review key terms by completing this crossword puzzle, using the following alphabetized list of terms:

adaptation
fossil
hominid
natural selection
Neanderthal
primate
taxonomy

Across
2 Common name for *Homo sapiens neanderthalensis*.
4 Group including australopithecines and humans.
5 Science of naming and classifying organisms.
6 Process by which populations become adapted to their environment.
7 Preserved remains of an organism.

Down
1 Animal that belongs to the order Primates.
3 Fitness of an organism for its environment.

CHAPTER TEST

OBJECTIVE TEST

Do not refer to the text when taking this test.

___ 1. Which of the following is/are used as evidence of the evolutionary process?
   a. fossils
   b. comparative anatomy
   c. comparative biochemistry
   d. All of these are correct.

___ 2. The vertebral bones that make up the remains of a human tail are considered to be
   a. vestigial organs.
   b. fossils.
   c. biochemical evidence for evolution.
   d. accidental.

___ 3. Which of these statements regarding the fossil record supports evolutionary theory?
   a. Chronologically speaking, fossils exhibit an increase in complexity of form.
   b. Fossils of prehistoric horses reflect a gradual change in form toward the modern horse.
   c. A number of intermediate forms have been found representing transitions between major taxonomic groups.
   d. Specialized features of fossilized organisms correlate well with what environments were believed to be like.
   e. All of these are correct.

____ 4. In the phrase *theory of evolution,* the term *theory* implies
   a. a general, unsupported notion.
   b. absolute fact.
   c. substantial evidence supports this idea.
   d. None of these is correct.
____ 5. Related organisms share
   a. similar anatomy.
   b. a common ancestor.
   c. genes.
   d. All of these are correct.
____ 6. The observation that most species overpopulate their environment best matches with which characteristic?
   a. members of a species vary
   b. struggle for existence
   c. survival of the fittest
   d. adaptation to the environment
____ 7. Which one of these is *not* a part of Darwin's theory of natural selection?
   a. Organisms strive to develop those traits that are most helpful in their particular environment.
   b. There is severe competition among organisms for available resources.
   c. The members of a species differ from each other.
   d. Better-adapted members of a species will survive and reproduce more successfully.
____ 8. The observation that whales have flippers best matches with which characteristic?
   a. members of a species vary
   b. struggle for existence
   c. survival of the fittest
   d. adaptation to the environment
____ 9. The observation that you can tell one domestic cat from another best matches with which characteristic?
   a. members of a species vary
   b. struggle for existence
   c. survival of the fittest
   d. adaptation to the environment
____ 10. The observation that female lions that are good mothers have the most surviving cubs best matches with which characteristic?
   a. Members of a species vary.
   b. Struggle for existence.
   c. Survival of the fittest.
   d. Adaptation to the environment.
____ 11. Which of these characteristics is best associated with natural selection?
   a. Organisms pass on their genes to offspring.
   b. Intelligence is a factor of brain size.
   c. The environment will determine who will reproduce.
   d. Overpopulation leads to inferior specimens.
____ 12. When life arose, which was not present?
   a. glucose
   b. water
   c. oxygen
   d. All of these are correct.
____ 13. The protocell was a(n)
   a. fermenter.
   b. autotroph.
   c. fermenting heterotroph.
   d. fermenting autotroph.
____ 14. Which pair does *not* give a proper sequence of events?
   a. heterotrophic nutrition–autotrophic nutrition
   b. bacteria evolve–plants evolve
   c. humans evolve–all other organisms evolve
   d. primitive gases–protocell
____ 15. Which of these does not describe the primitive atmosphere?
   a. ice cold
   b. no free oxygen
   c. contained gases
   d. no flying creatures
____ 16. A modern ape, like a human, is a
   a. mammal.
   b. primate.
   c. hominoid.
   d. All of these are correct.
____ 17. Which of these are modern humans?
   a. Neanderthals
   b. *Homo erectus*
   c. *Homo habilis*
   d. Cro-Magnon
____ 18. Which of these are hominids?
   a. great apes
   b. australopithecines
   c. humans
   d. Both *b* and *c* are correct.
____ 19. *Homo habilis* is a(n)
   a. prosimian.
   b. australopithecine.
   c. human.
   d. All of these are correct.
____ 20. Which is true of humans but not of apes?
   a. flexible spine
   b. arms shorter than legs
   c. walking upright
   d. All of these are correct.
____ 21. Which of these is paired with an inappropriate comment?
   a. *Australopithecus africanus*–could be direct ancestor to humans
   b. *Australopithecus robustus*–probably hunted large game
   c. *Australopithecus afarensis*–had a small brain but walked erect
   d. All of these are correct.

____22. Which of these is paired with an inappropriate comment?
   a. *Homo habilis*–tool maker
   b. *Homo erectus*–big-game hunter
   c. *Homo sapiens neanderthalis*–remains found only in Africa
   d. Cro-Magnon–drew beautiful pictures

____23. *Homo erectus* is believed to be the hominid closest to humans because
   a. the cranial capacity was nearly that of modern humans.
   b. these hominids used tools.
   c. these hominids were bipedal.
   d. the opening for the spinal cord was more forward on the skull than in other fossils.
   e. these hominids cultivated crops.

____24. Which of these describes the climate when humans began evolving?
   a. The hot and humid climate produced thick jungles.
   b. Trees were disappearing, and grasslands were more common.
   c. It was so dry that much of Africa became like the Sahara.
   d. This was the time of the first ice age.

____25. Evidence suggests that
   a. we are descended from one of the modern-day great apes.
   b. we couldn't possibly be related to apes.
   c. we share a common ancestor with modern-day apes.
   d. humans evolved in Africa and possibly China.
   e. Both *c* and *d* are correct.

THOUGHT QUESTIONS

Answer in complete sentences.

26. Contrast Darwin's emphasis on *survival of the fittest* with the modern emphasis on *differential reproduction* to explain the occurrence of natural selection.

27. How might big-game hunting have influenced the social organization of humans?

Test Results: _____ number correct ÷ 27 = _____ × 100 = _____%

Answer Key

Study Questions

1. a. embryology **b.** biochemical **c.** homologous structures **d.** history of life **e.** geography **2. a.** fossil **b.** 100,000 **c.** DNA, ATP, DNA triplet code **d.** vestigial **3.** It indicates there is substantial evidence to support the idea of evolution. **4.** d, b, e, c, a **5. a.** no **b.** their anatomy and/or physiology are different **6.** There are more members than can be comfortably supported by the environment. **7.** a **8.** b **9. a.** survived **b.** reproduce more successfully **10. a.** primitive atmosphere contains gases **b.** simple organic molecules form **c.** macromolecules lead to protocell **d.** oxygen in atmosphere and aerobic respiration evolves **11. a.** oxygen **b.** water **c.** simple organic molecules **d.** energy sources **e.** organic soup **f.** macromolecules **g.** heterotrophic fermenter **12. a.** For example, "Karen pushed cans off friendly grandmother's stove." (kingdom, phylum, class, order, family, genus, species) **b.** genus first, then species **c.** No, taxonomy changes as more information is gathered about evolutionary relationships. **13. a.** Animalia **b.** Chordata **c.** Mammalia **d.** Primates **e.** Hominidae **f.** *Homo* **g.** *sapiens* **14. a.** opposable thumb **b.** expanded cerebrum **c.** living in trees **d.** expanded cerebrum **e.** long childhood **15. a.** cats, dogs, horses, mice, sheep, cattle **b.** monkeys, apes, humans **c.** extinct humans (*Australopithecus, Homo habilis, Homo erectus, Homo sapiens neanderthalensis*) and modern humans (*Homo sapiens*) **16. a.** australopithecines, walking erect **b.** *Homo habilis*, tool use **c.** *Homo erectus*, knowledge of fire **d.** Neanderthal, *Homo sapiens* **e.** Cro-Magnon, modern appearance **17.** Perhaps the Neanderthals evolved into Cro-Magnon or interbred with them. **18. a.** Molecular data show that there is just as much variation within races as there is between races. No definite molecular pattern of racial characteristics is evident. **b.** climate

Definitions Crossword

Across
2 Neanderthal 4 hominid 5 taxonomy 6 natural selection 7 fossil

Down
1 primate 3 adaptation

Chapter Test

1. d **2.** a **3.** e **4.** c **5.** d **6.** b **7.** a **8.** d **9.** a **10.** c **11.** c **12.** c **13.** c **14.** c **15.** a **16.** d **17.** d **18.** d **19.** c **20.** d **21.** b **22.** c **23.** a **24.** b **25.** e **26.** Darwin emphasized that the most fit members of a species survived while the least fit died off, but modern evolutionists emphasize that the most fit members of a species reproduce to a greater degree than the least fit members. **27.** Big-game hunting seems to have increased social cooperation among humans. Communication between members of the group developed, and the men brought the kill home to be shared by all. Big-game hunting may account for division of labor between the sexes. Men went out to hunt, while women stayed home to nurture young.

23

ECOSYSTEMS

STUDY TIPS

In chapter 22, you learned that humans are the product of a long evolutionary history, closely related to other animals, especially the primates. In this chapter, you will learn that humans are a part of natural ecosystems. The biological principles that govern natural ecosystems also apply to us. We have disregarded many of those principles, however, in our quest to possess and control more of the earth.

The flow of energy is a one-way process through a community of organisms. Many food chains can be interwoven into a food web (p. 481), and much energy is lost at each trophic level (p. 481) within them. This energy relationship forms the basis for ecological pyramids of numbers or biomass (pp. 481–482). Write or draw an example of a food chain, one of a food web, and another of an ecological pyramid.

Learn the basic diagram for biogeochemical cycles (p. 482) first, and then add specifics for each nutrient cycle (phosphorus, p. 484; nitrogen, pp. 485–486; and carbon, p. 487). Indicate how humans have affected these cycles with their various activities.

STUDY QUESTIONS

Study the text section by section as you answer the questions that follow.

23.1 THE NATURE OF ECOSYSTEMS (PAGES 476–482)

- Discuss the concepts of succession and climax community.
- Understand that natural communities are dynamic and interact with the environment.
- Discuss ecosystem composition and the concept of niche, and provide examples.

1. Match the terms in the key with the following definitions.
    Key *biosphere   communit   ecology   ecosystem   population*

    _____ a. A community plus its physical habitat.

    _____ b. Members of the same species in the same area.

    _____ c. Study of the interactions of organisms with each other and their surroundings.

    _____ d. The living layer surrounding the earth.

    _____ e. All the populations interacting in an area.

2. Place the following items in the proper order to describe secondary succession: *mature forest, annual weeds, shrubs.*
   First, a._____; then, b._____; then, c._____.

3. List the components of the niche of a producer. _____
   _____

4. Study the following diagram, and answer these questions.
   Why does a wide arrow point from both producers and consumers to decomposers? a._____
   _____

   Why does a wide arrow point only from inorganic nutrient pool to producers? b._____
   _____

   Why don't the wavy arrows point from decomposers to producers? c._____
   _____

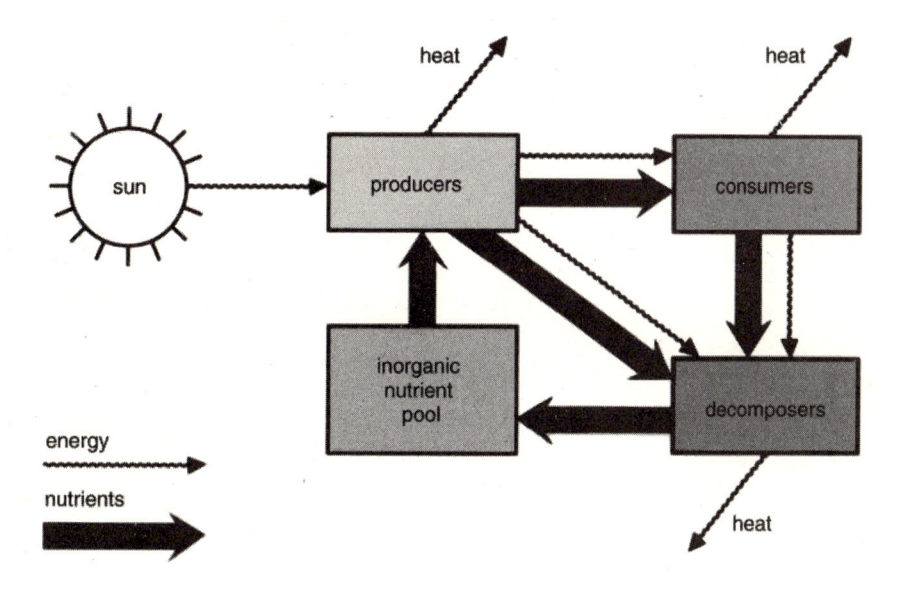

197

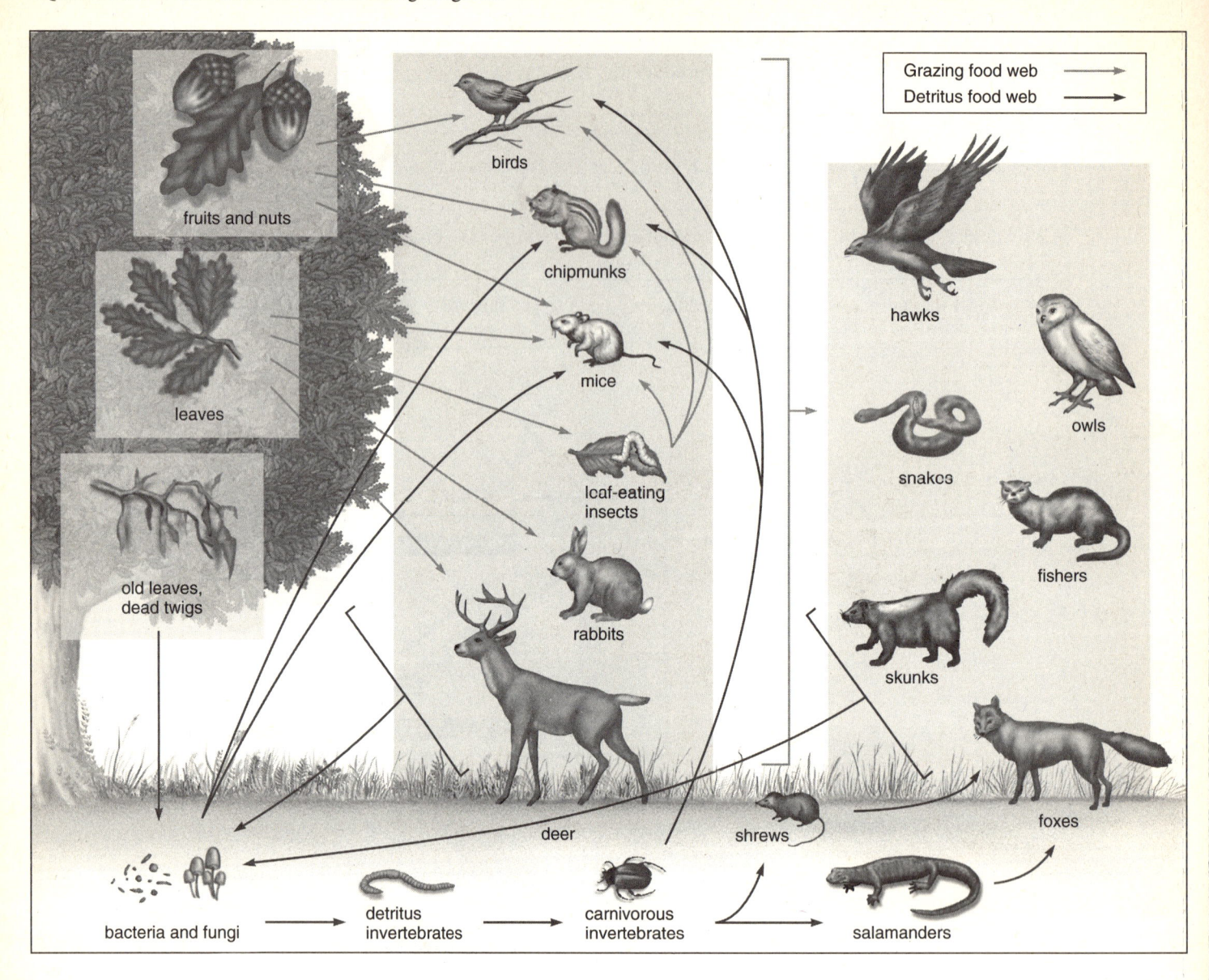

Questions 5–7 are based on the following diagram:

5. From this food web, formulate a grazing food chain.

6. From this food web, formulate a detritus food chain.

7. Explain one way in which the detritus food web and the grazing food web are always connected.

198

Questions 8–9 are based on the following diagram:

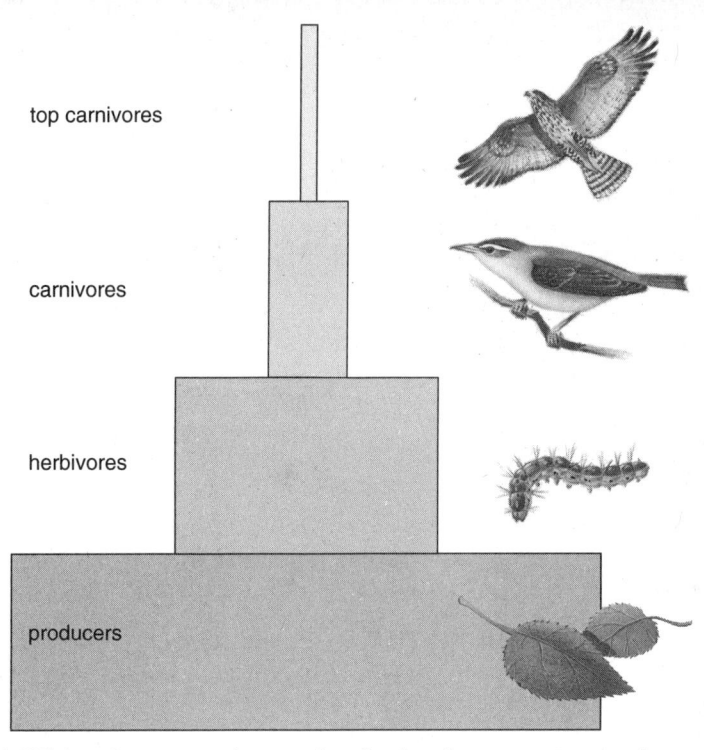

top carnivores

carnivores

herbivores

producers

8. Each level of a pyramid represents a trophic level. With reference to the grazing food web, name an organism at the first trophic level. a. _____ Two organisms at the second trophic level.
b. _____ Two organisms at the third trophic level. c. _____

9. With reference to a pyrmid of energy, why is each trophic level reduced in size? a. _____
Why is there a limited number of trophic levels? b. _____

23.2 GLOBAL BIOGEOCHEMICAL CYCLES (PAGES 482–487)

- Biogeochemical cycles are gaseous (carbon cycle, nitrogen cycle) or sedimentary (phosphorus cycle)
- The addition of carbon dioxide (and other gases) to the atmosphere is associated with global warming.
- The production of fertilizers from nitrogen gas is associated with acid deposition, photochemical smog, and temperature inversions.
- Fertilizer also contains mined phosphate; fertilizer runoff is associated with water pollution.

10. Examine the following diagram and then answer the questions:

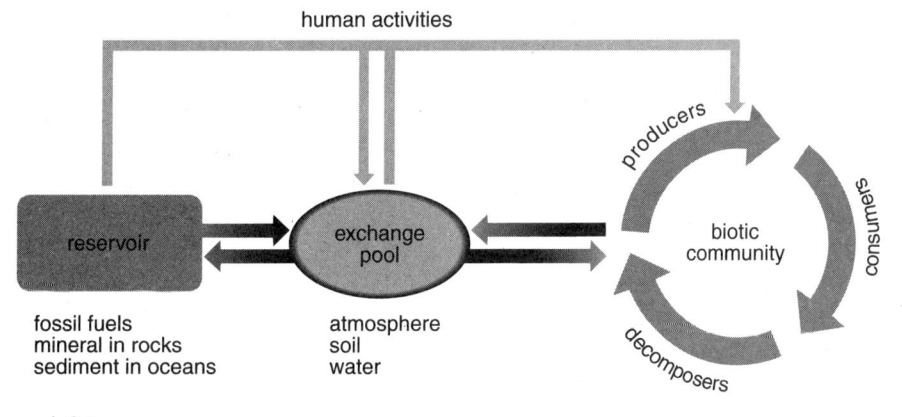

human activities

reservoir → exchange pool → biotic community (producers, consumers, decomposers)

fossil fuels
mineral in rocks
sediment in oceans

atmosphere
soil
water

What is a reservoir? a. _____
What is an exchange pool? b. _____
What is a biotic community? c. _____
Explain the arrows labeled *human activities*. d. _____
_____

199

The Water Cycle (Page 483)

11. Complete this diagram of the water cycle by: filling in the boxes, using these terms: *ice, H₂O in the atmosphere, ocean, groundwaters*
    labeling the arrows, using these terms: *precipitation* (twice), *transpiration from plants and evaporation from soil, evaporation, transport of water vapor by wind.*

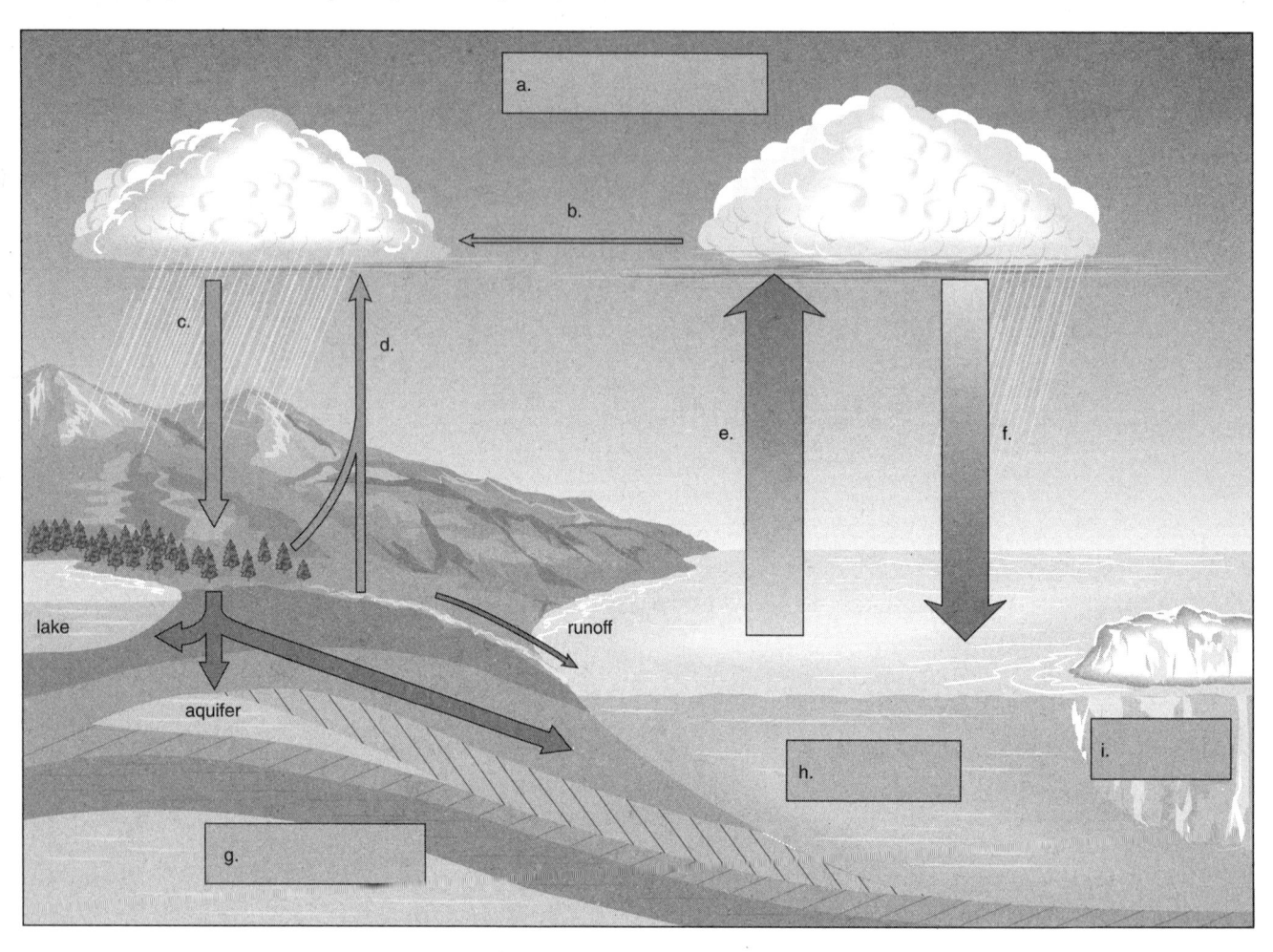

12. Select the statements that are true about the water cycle.
    _____ a. Water cycles between the land, the atmosphere, and the ocean and vice versa.
    _____ b. We could run out of fresh water.
    _____ c. The ocean receives more precipitation than the land.
    _____ d. Water that is in the aquifers never reaches the oceans.

The Phosphorus Cycle (Page 484)

13. Place a check in front of the statement(s) that describe(s) the results when producers take up phosphate.
    _____ a. become a part of phospholipids
    _____ b. becomes a part of ATP
    _____ c. becomes a part of nucleotides
    _____ d. become a part of the atmosphere

14. Indicate whether the statements that follow are true (T) or false (F). Rewrite all false statements to be true statements.
    _____ a. Excess phosphate in bodies of water may cause radiation poisoning. Rewrite: _____
    _____
    _____ b. Most ecosystems have plenty of phosphate. Rewrite: _____
    _____

200

_____ c. The phosphorus cycle is a sedimentary cycle. Rewrite: _____
_____

_____ d. Phosphate enters ecosystems by being taken up by animals. Rewrite: _____
_____

The Nitrogen Cycle (Pages 485–486)

Questions 15 and 16 are based on the following diagram:

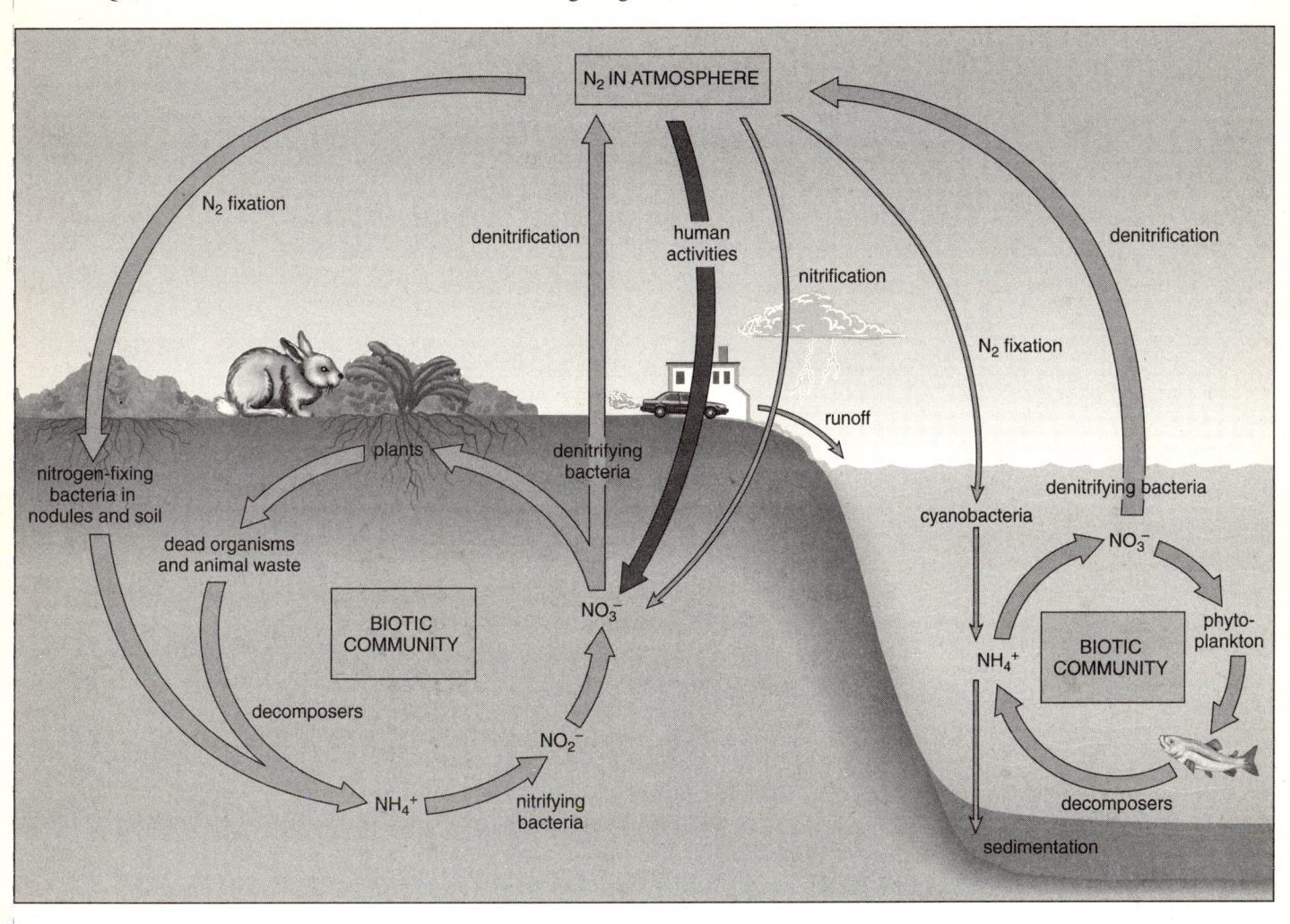

15. Match the definitions with the following terms:
    1 denitrifying bacteria
    2 nitrifying bacteria
    3 nitrogen-fixing bacteria
    _____ a. bacteria that convert nitrate to nitrogen gas.
    _____ b. bacteria that convert ammonium to nitrate
    _____ c. bacteria in legume nodules that convert nitrogen gas to ammonium
16. Plants cannot utilize nitrogen gas. What are two ways in which plants receive a supply of nitrogen for incorporation into proteins and nucleic acids? _____

The Carbon Cycle (Page 487)

17. In the carbon cycle, carbon dioxide is removed from the atmosphere by the process of a. _____ but is returned to the atmosphere by the process of b. _____ Living things and dead matter in soil are carbon c. _____ and so are the d. _____ because of shell accumulation. In aquatic ecosystems, carbon dioxide from the air combines with water to produce e. _____ that algae can use for photosynthesis. In what way do humans alter the exchange rates in the carbon cycle? f. _____

201

23.3 HUMAN-IMPACTED ECOSYSTEMS (PAGES 488–490)

- Describe human-impacted ecosystems.
- Suggest ways in which the human ecosystem could become more cyclical.

18. Name two major inputs and two outputs of human-impacted ecosystems.

    Inputs   Outputs

    a._____   c._____

    b._____   d._____

19. The following variables contribute to the overall impact humans have on natural ecosystems. Indicate whether they occur in the country or in the city or both.

    _____ a. excessive fuel consumption

    _____ b. heavy use of fertilizers, pesticides, herbicides

    _____ c. excessive human sewage (pollution)

    _____ d. generous irrigation

20. List three ways that conservation of resources can be achieved.

    a._____

    b._____

    c._____

DEFINITIONS CROSSWORD

Review key terms by completing this crossword puzzle, using the following alphabetized list of terms:

consumer
detritus
ecology
ecosystem
food chain
niche
omnivore
producer
succession

Across
1. The study of the interactions of organisms with each other and their habitat.
5. Animal that feeds on both plants and animals.
6. Series of ecological stages until a climax is reached.
7. Role of an organism within its community.
8. Organism at the start of the food chain.

Down
1. Community plus its nonliving habitat.
2. Series of organisms linked together by feeding relationships.
3. Nonliving organic matter.
4. Organism that feeds on another living organism.

Chapter Test

Objective Test

Do not refer to the text when taking this test.
For questions 1–4, use this key.

Key a. ecology   b. population   c. community
   d. ecosystem

____ 1. a community plus its physical environment
____ 2. study of the interactions of organisms with each other and with the physical environment
____ 3. all populations that are found in a particular area
____ 4. all members of the same species that inhabit a particular area
____ 5. An example of primary succession is
   a. aquatic plants growing along the banks of an ancient river.
   b. annual weeds growing on an abandoned agricultural field.
   c. annual weeds starting on a new sandbar.
   d. shrubs and small trees at the edge of the forest.
____ 6. An example of secondary succession is
   a. aquatic plants growing along the banks of an ancient river.
   b. annual weeds growing on an abandoned agricultural field.
   c. annual weeds starting on a new sandbar.
   d. shrubs and small trees at the edge of the forest.
____ 7. Which of these represents a climax community?
   a. the desert of the southwestern United States
   b. midcontinent grasslands
   c. temperate rain forest
   d. all of these
____ 8. The _____ of an organism is its profession or total role within the community.
   a. habitat
   b. ecology
   c. trophic level
   d. niche
____ 9. Which of the following organisms is considered a producer?
   a. fungus
   b. earthworm
   c. algae
   d. bumblebee

For questions 10–14, use this key.
   Key a. energy   b. chemicals   c. producer
      d. carnivore   e. food web

____10. a continual outside supply needed
____11. cycle through an ecosystem
____12. the first member of any food chain
____13. the intertwining of many food chains
____14. always at the top of the food pyramid

Questions 15 and 16 refer to this stable food chain:
   grass—rabbits—snakes—hawks

____15. Each population is
   a. larger than the one before.
   b. an appropriate size to support the next.
   c. a herbivore.
   d. All of these are correct.
____16. Rabbits are
   a. consumers.
   b. herbivores.
   c. more plentiful than snakes.
   d. All of these are correct.

Questions 17 and 18 refer to this food chain:
   corn—pigs—humans

____17. This food chain (in comparison with the preceding one) is
   a. better.
   b. not common.
   c. not wasteful.
   d. simpler.
____18. If crows were eating the corn seed and farmers killed them,
   a. this would have no effect on natural ecosystems.
   b. this might disrupt a natural ecosystem.
   c. it would tend to stabilize ecosystems.
   d. the cornfield would disappear.
____19. Which of these statements is correct?
   a. The base of the energy pyramid contains the largest trophic level.
   b. About 10% of energy available in food is actually incorporated into any trophic level.
   c. Humans are at the top of energy pyramids.
   d. all of these
____20. Which of the following contributes to the carbon cycle?
   a. respiration
   b. photosynthesis
   c. fossil fuel combustion
   d. all of these
____21. The carbon exchange pool is in
   a. the soil.
   b. the atmosphere.
   c. the ocean.
   d. deep sediments.
____22. Which group of organisms is directly responsible for the release of the majority of inorganic nutrients from organic substances?
   a. fungi and bacteria
   b. small arthropods
   c. earthworms

___ 23. Which of the following is *not* a type of bacteria in the nitrogen cycle?
   a. nitrogen-fixing
   b. nitrifying
   c. nitrosamines
   d. denitrifying
   e. decomposers

___ 24. The form of nitrogen most plants make use of is
   a. atmospheric nitrogen.
   b. nitrogen gas.
   c. organic nitrogen.
   d. nitrates.

___ 25. The human-impacted ecosystem
   a. is largely noncyclical.
   b. uses too much energy input.
   c. is dependent on the natural ecosystems.
   d. all of these

THOUGHT QUESTIONS

Answer in complete sentences.

26. How will the burning of tropical rain forests contribute to the carbon cycle?

27. What type of plant should farmers plant to increase the nitrogen content of the soil? Why?

Test Results: _____ number correct ÷ 27 = _____ × 100 = _____%

ANSWER KEY

STUDY QUESTIONS

1. a. ecosystem **b.** population **c.** ecology **d.** biosphere **e.** community **2. a.** annual weeds **b.** shrubs **c.** mature forest **3.** Components of the niche include the different organisms that live on its body, the consumers that use it as food, how it cycles nutrients, how much space it occupies within its ecosystem, how much detritus it contributes to the soil each year, and so forth. **4. a.** Both producers and consumers die and decompose. **b.** Only producers use inorganic nutrients to make organic molecules. **c.** Energy does not cycle. **5.** tree leaves, rabbits, hawks **6.** dead twigs, bacteria and fungi, detritus invertebrates, carnivouous invertebrates, shrews, foxes **7.** Members of the grazing food web die and are decomposed by bacteria and fungi. **8. a.** trees **b.** deer and rabbits **c.** snakes and hawks **9. a.** As energy flows from one trophic level to the next, approximately 90% of it is lost. **b.** Eventually there is not enough energy to support another level. **10. a.** A source of a chemical that is generally unavailabe for a biotic community. **b.** A source of the chemical that is availabe to a biotic community. **c.** The populations that interact with each other and the abiotic environment within an ecosystem. **d.** Humans increase the transfer of chemicals from reservoirs and from exchangeand from reservoirs and from exchange pools so that they are then available to the biotic community at a higher rate. **11. a.** H_2O in atmosphere **b.** transport of water vapor by wind **c.** precipitation **d.** transpiration from plants and evaporation from soil **e.** evaporation from ocean **f.** precipitation **g.** ground waters **h.** ocean **i.** ice **12.** a, b, c **13.** a, b, c **14. a.** F, . . . may cause algal bloom **b.** F, . . . have a limited supply of phosphate **c.** T **d.** F, . . . taken up by plants. **15. a.** 1 **b.** 2 **c.** 3 **16.** nitrogen-fixing bacteria in nodules and nitrate in soil **17. a.** photosynthesis **b.** cellular respiration **c.** reservoirs **d.** oceans **e.** bicarbonate **f.** by burning fossil fuels that add carbon to the atmosphere **18. a.** fuel energy **b.** raw materials **c.** waste **d.** pollution **19. a.** both **b.** country **c.** city **d.** country **20. a.** wise use of only amounts needed **b.** recycling **c.** use of renewable energy sources

DEFINITIONS CROSSWORD

Across
1 ecology 5 omnivore 6 succession 7 niche 8 producer
Down
1 ecosystem 2 food chain 3 detritus 4 consumer

CHAPTER TEST

1. d **2.** a **3.** c **4.** b **5.** c **6.** b **7.** d **8.** d **9.** c **10.** a **11.** b **12.** c **13.** e **14.** d **15.** b **16.** d **17.** d **18.** b **19.** d **20.** d **21.** b **22.** a **23.** c **24.** d **25.** d **26.** The burning of tropical rain forests will release carbon dioxide into the atmosphere and increase the greenhouse effect. **27.** Farmers should plant legumes (e.g., soybeans) to increase the nitrogen content of the soils because these plants have root nodules where bacteria fix atmospheric nitrogen. Rotating with other crops is a wise use of agricultural land.

24

POPULATION CONCERNS

STUDY TIPS

In chapter 23, you saw how humans are a part of the natural community and how many of our activities are disruptive to the workings of natural ecosystems. Since we are able to artificially produce a comparatively predictable food supply rather than gather it from the wild, our numbers have increased. Much more so, in fact, than we would have if we continued as hunter-gatherers and our numbers were contingent upon the natural food supply and environment.

Make up several problems to help you use the equation for population growth (pp. 496–497). Understand what is meant by carrying capacity (p. 497) and the result of exceeding that capacity. Make a list of the population traits of less-developed and more-developed countries (p. 498) to help you remember them. Similarly, make a flowchart showing the types of air pollutants, their causes, and their effects (pp. 500–504). Understand that how we use water (pp. 505–506) contributes to water pollution, and note the realized and potential effects of global deforestation (p. 508) on our fragile planet. Lastly, make a list of ways you and your friends and family can make adjustments in your life-styles to lessen your personal impact on the environment. Try your best to implement your list.

STUDY QUESTIONS

Study the text section by section as you answer the questions that follow.

24.1 HUMAN POPULATION GROWTH (PAGES 496–499)

- Understand the basic principles of exponential growth.
- Calculate the growth rate when given the birthrate and death rate.
- Calculate the doubling time when given the growth rate.
- Describe population size as a compromise between biotic potential and environmental resistance.

1. Suppose a population of 5,000 persons had a yearly increase of 10%. What would be the population size after two years? a._____

   Why is the second year's increase greater than the first year's increase? b._____

   c. Which of these statements comes closest to describing the world's current population?
   The world population increases by the number of people found in a medium-sized city (200,000) _____
   every year.
   every six months.
   every month.
   every day.

2. What is the growth rate of a population in which the birthrate is 40/1,000 and the death rate is 12/1,000? ___

3. What is the doubling time of the population referred to in question 2? a._____

   What is the doubling time of the world's population today? b._____

   There were about 5 billion people alive in 1984. In what year will there be 9 billion, assuming no increase or decrease in the growth rate? c._____

4. Label the following diagram using the alphabetized list of terms. Note: You will use all four terms; two of them appear at the same point on the diagram.
   biotic potential
   carrying capacity
   environmental resistance
   exponential growth

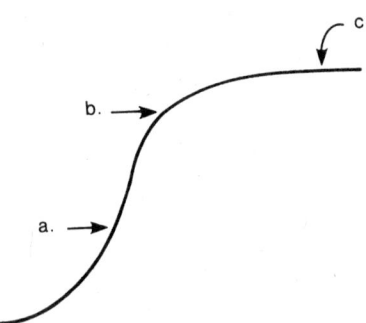

In questions 5–7, indicate whether the following statements are true (T) or false (F). Rewrite the false statements to make true statements.

5. _____ a. The demographic transition is accompanied by an increase in growth rate. Rewrite:

   _____ b. The more-developed countries have not undergone the demographic transition. Rewrite:

   _____ c. Great Britain is an example of a less-developed country. Rewrite:

6. _____ The birthrate has declined in less-developed countries but not the death rate. Rewrite:

7. _____ The age structure diagram for less-developed countries has the appearance of a pyramid because postreproductive citizens outnumber the prereproductive citizens. Rewrite:

8. Even though the growth rate is declining, the populations of countries in the areas of
   a._____, b._____, and c._____
   are still expanding because of d._____ growth.

9. What three ways have been suggested to help reduce the growth rate in the less-developed countries?
   a._____
   b._____
   c._____

24.2 THE HUMAN POPULATION AND POLLUTION (PAGES 500–508)

- Discuss three problems associated with air pollution.
- Discuss three problems associated with water pollution.
- Discuss three problems associated with land degradation.

Match each gas in questions 10–14 to a source and an air pollutant effect.

Source

a. vehicle exhaust
b. fossil fuel burning
c. paint solvent
d. pesticides
e. deforestation
f. farms
g. refrigerants
h. fire extinguishers
i. plastic foam

Air Pollutant Effect

j. photochemical smog
k. acid deposition
l. global warming
m. ozone shield destruction

10. NO$_x$ _____ and _____
11. CFCs _____ and _____
12. HC _____ and _____
13. CH$_4$ _____ and _____
14. SO$_2$ _____ and _____

15. What is the effect of each of the following?
    excess nutrients in bodies of water a. _____

    organochlorides in drinking water b. _____

    aluminum and iron leached from soil by acid rain c. _____

16. Aquifers are polluted when a. _____ percolates into the underground water supply. Chemical wastes are sometimes b. _____ into deep wells or dumped into the c. _____ . Sometimes, these wastes can be prevented by d. _____. In any case, the ocean is the final resting place for all types of water e. _____.

17. Indicate whether the following statements about biological magnification are true (T) or false (F):
    _____ a. Posisons such as DDT decrease as they pass upward from one trophic level to another.
    _____ b. Biological magnification is more likely in aquatic food chains.
    _____ c. Humans are not affected by biological magnification.
    _____ d. Posions like DDT become more magnified (concentrated) as they pass upword from one trophic level to anothe.

18. Complete the following phrases to explain how land is being degrades today:

    Soil erosion causes a loss of a. _____

    Desertification means b. _____

    Forests are c. _____

19. Explain how forest destruction causes each of the following:

    loss of a CO$_2$ sink a. _____

    loss of biodiversity b. _____

    loss of medicinal plants c. _____

    soil erosion d. _____

    water pollution e. _____

    ecosystem destruction f. _____

24.3 THE HUMAN POPULATION AND BIODIVERSITY (PAGES 509–510)

- Conservation biology is the scientific study of biodiversity and ecosystem management for the preservation of all species, including *Homo sapiens*.

20. What three levels interplay to make up biodiversity? _____

21. List four human activities that lead to the extinction of other organisms. _____

22. What new area of biology is concerned with the preservation and management of ecosystems for sustainability? _____

DEFINITIONS WORDSEARCH

Review key terms by completing this wordsearch, using the following alphabetized list of terms:

```
O B I O T I C P O T E N T I A L N
T Z E D E S U O H N E E R G R A V
G S O F E C U N D L I B R R R Y D
N U B N Y A N D R G E A T O L O K
I S A X E L A T J U L P O W P R Y
Y T E R W S Q Y P O L L U T A N T
R A G B A S H V Y J U L P H A N T
R I B S A Q U I F E R F O R A R T
A N F M A I E M E O F D O A F A R
C A A W H D O U B L I N G T I M E
P B T O M E W H A T D D I E H U N
I L T F O R D E R O C T B E O R B
L E N V I R O N M E N T A L K I L
```

aquifer
biotic potential
carrying
doubling time
environmental
greenhouse
growth rate
ozone shield
pollutant
sustainable

a. _____ Number of years it takes to double population size.
b. _____ Shield formed from oxygen in the upper atmosphere.
c. _____ Undesirable substance added to environment.
d. _____ Underground water reservoir.
e. _____ Maximum population growth under ideal conditions.
f. _____ Numerical capacity of a species that can be supported by the area indefinitely.
g. _____ Effect of reradiation of solar heat due to gases in the atmosphere.
h. _____ The yearly percentage of increase or decrease in the size of a population.
i. _____ Global way of life that recycles, conserves, protects.
j. _____ Type of resistance due to factors that limit the increase of a population's size.

Notes

Notes

Notes

Notes

Notes

Notes

Notes

Notes